THE SURVEILLANCE STATE

A Novel

Geoffrey Arbuckle

CANE Enterprises Publishing, LLC

This is a work of fiction. Names, characters, businesses, places, events, locales, and incidents are either the products of the author's imagination or used in a fictitious manner. Any resemblance to actual persons, living or dead, or actual events is purely coincidental.

Every technology described in this book exists or is in active development.

Published by CANE Enterprises Publishing, LLC

Kansas City, Missouri

We all clicked "I Agree."

PROLOGUE — "TERMS AND CONDITIONS"

Before you read this book, you should know how we got here. Not to the fictional city of this novel — to the real one. To the world outside your window. To the phone in your hand.

Every fact in this prologue is true. Every technology is real. Every law was passed. Every breach occurred. Every product was purchased. The only fiction begins in Chapter 1.

* * *

I. The Architecture (1994-2006)

In 1994, a programmer named Lou Montulli invented the HTTP cookie — a small text file that allowed a website to remember your browser. The cookie was designed to solve a practical problem: online shopping carts kept forgetting what you'd put in them. The solution was a tiny piece of code that let the website remember you. The remembering was the beginning. Before the cookie, the internet forgot you the moment you left. After the cookie, the internet never forgot you again.

By 1998, the advertising company DoubleClick had deployed cookies across thousands of websites, building profiles of individual users' browsing habits — what they read, what they searched for, what they bought, how long they lingered on a page. The profiles were anonymous. They were also comprehensive. A user who searched for pregnancy tests and baby furniture and OB-GYN reviews was, to the cookie, a data pattern. The pattern was worth money. DoubleClick sold the patterns to advertisers. The advertisers targeted the patterns. The patterns became the product. You became the product. This was 1998. The iPhone didn't exist yet.

Google was three months old. Mark Zuckerberg was fourteen.

On September 12, 2001 — one day after the attacks — the NSA began collecting domestic phone records in bulk under a program that would eventually be called Stellar Wind. The legal authority was the Authorization for Use of Military Force, passed by Congress the following week with a single dissenting vote. The program collected the metadata — not the content — of every domestic phone call: who called whom, when, for how long, from where. The distinction between metadata and content was supposed to be a safeguard. It was a fiction. Metadata told you who a person called at 2 a.m. It told you the call lasted forty-five minutes. It told you the next call was to a divorce attorney. It told you the story without telling you the words, and the story was enough.

The Patriot Act, signed October 26, 2001, expanded the government's surveillance authority to include business records, library records, and "any tangible thing" relevant to a terrorism investigation. Section 215 became the legal foundation for the bulk collection of phone metadata that the NSA would conduct for fourteen years before a single member of the public knew it existed. The law was renewed in 2005 and 2010 and 2011. Each renewal was bipartisan. Each renewal expanded the authority. The surveillance state was not built by one party. It was built by both, in the name of a threat that was real and a response that was permanent.

***

II. The Device (2007-2013)

On June 29, 2007, Apple released the iPhone. The device contained a camera, a microphone, a GPS receiver, an accelerometer, and a cellular radio that transmitted its location to the nearest cell tower every twelve seconds. Steve Jobs called it "a revolutionary product that changes everything." He was right. The iPhone was the first surveillance device that people stood in line to buy. It was beautiful. It was intuitive. It fit in your pocket. It knew where you were at all times and it told Apple and AT&T and every app you installed, and you agreed to all of it when you tapped the green button on the Terms and Conditions screen that was 47 pages long and that you did not read because nobody read them because the phone was right there and it was beautiful and you wanted it.

By 2010, there were 300 million smartphones in the world. By 2012, one billion. Each one a camera. Each one a microphone. Each one a GPS tracker. Each one transmitting its owner's location, contacts, messages, photos, searches, browsing history, and biometric data to servers controlled by companies whose business model was the collection and monetization of that data. The smartphone was the most successful surveillance deployment in human history, and it was accomplished not by a government mandate but by a consumer product launch.

In 2009, a company called Waze released a navigation app that used its users' phones to track traffic patterns in real time. The app was free. The product was the drivers' location data — millions of phones reporting their speed and position every second, the aggregate forming a real-time map of every road in every city. Google acquired Waze in 2013 for $1.15 billion. The price was not for the app. The price was for the data — the location histories of 50 million people, updated continuously, stored indefinitely.

On June 5, 2013, The Guardian published the first of the Edward Snowden revelations: the NSA was collecting phone records of millions of Verizon customers under a secret court order. Over the following weeks, Snowden revealed PRISM — a program that gave the NSA direct access to the servers of Google, Apple, Facebook, Microsoft, Yahoo, and other tech companies. He revealed XKeyscore — a system that allowed analysts to search, without a warrant, the content of emails, chats, and browsing history of any person on earth. He revealed that the NSA had tapped the fiber optic cables that carried the world's internet traffic and was copying the data as it passed through.

The revelations were the largest leak of classified intelligence in American history. They produced global outrage, congressional hearings, a presidential review panel, and the USA Freedom Act of 2015, which placed modest restrictions on bulk metadata collection. The NSA's capabilities were not dismantled. They were refined. The commercial surveillance infrastructure — the smartphones, the apps, the cookies, the data brokers — continued to expand, unaffected, because the commercial infrastructure was not covered by the reforms. The government's ability to collect your data was slightly curtailed. The corporations' ability to collect the same data and sell it to the government was untouched.

Nothing changed. The architecture held.

* * *

III. The Home (2014-2019)

On November 6, 2014, Amazon released the Echo — a cylindrical speaker with a microphone array that listened for the wake word "Alexa" and responded to voice commands. The Echo was the first always-listening device designed for the living room. Amazon sold it as a convenience: play music, set timers, check the weather. The microphone array contained seven directional microphones capable of detecting a whisper from across a room. The device was always on. The device was always listening. Amazon's privacy policy stated that audio was recorded only after the wake word was detected. Reports later revealed that the Echo recorded and stored audio snippets without the wake word, that Amazon employees listened to recordings to "improve the service," and that Amazon had provided Echo recordings to law enforcement in at least one criminal case.

The Echo was the beginning of the smart home — the transformation of the domestic space from a private refuge into a networked sensor array. By 2018, the market included: Amazon Echo and Echo Show (microphones and cameras in the kitchen, the bedroom, the bathroom). Google Nest Hub (cameras and microphones). Ring doorbell (cameras on the front door, footage stored on Amazon's servers, shared with over 2,000 law enforcement agencies through Amazon's Neighbors Public Safety program without requiring a warrant). Nest thermostat (occupancy sensors that knew when you were home). Smart TVs from Samsung and LG (microphones that listened for voice commands and, according to Samsung's own privacy policy, captured "personal or other sensitive information" that was "among the data captured and transmitted to a third party through voice recognition"). Robot vacuums from iRobot (mapping every room, every hallway, every piece of furniture — the floor plan uploaded to the cloud). Smart mattresses from Sleep Number (tracking heart rate, respiratory rate, and movement throughout the night — the data shared with "business partners").

Each device was sold as a convenience. Each device was a sensor. The sensors were connected to the cloud. The cloud was connected to the companies. The companies were connected to the data brokers, the advertisers, the analytics firms, the insurance companies, and the law enforcement agencies. The home that had once been the most private space in American life was now the most surveilled, and the surveillance had been invited in by the homeowner, who had purchased it at Best Buy and plugged it into the wall and said "Alexa, play jazz."

* * *

IV. The Face (2016-2023)

In 2016, a company called Clearview AI scraped three billion photographs from social media platforms — Facebook, Instagram, Twitter, YouTube — and used them to build a facial recognition database available to law enforcement. The database allowed police to photograph a face and match it against three billion images in seconds. Clearview AI did not have permission from the people in the photos. It did not have permission from the platforms. It did not need permission, because the photos had been posted publicly, and publicly posted photos were not protected by any federal privacy law.

By 2020, Clearview AI's database had been used by over 600 law enforcement agencies in the United States. Studies by MIT and the National Institute of Standards and Technology found that facial recognition systems misidentified Black and Asian faces at rates five to ten times higher than white faces. At least three men — all Black — were wrongfully arrested based on facial recognition misidentification between 2019 and 2023. The technology was deployed fastest in the neighborhoods where it was least accurate.

In 2020, the COVID-19 pandemic normalized location surveillance on a global scale. Contact tracing apps — endorsed by Apple and Google, deployed by governments — tracked the proximity of every phone to every other phone, logging encounters, flagging exposures, mapping the movement of populations in real time. The apps were voluntary. The infrastructure was permanent. When the pandemic ended, the contact tracing apps were retired. The location tracking infrastructure was not.

In 2021, a security breach at Verkada — a company that provided cloud-connected cameras to hospitals, prisons, schools, and Tesla factories — exposed live feeds from over 150,000 cameras. The breach revealed that Verkada's cameras in schools were equipped with facial recognition that tracked individual students. The cameras in hospitals captured patients in psychiatric wards. The cameras in prisons monitored inmates in their cells. The footage was stored in the cloud and accessible to any Verkada employee with the right credentials. The breach was discovered by a Swiss hacker who said the point was to show "how pervasive video surveillance has become."

* * *

V. The Body (2023-2026)

In 2023, Apple released the Vision Pro — the first mass-market spatial computing headset with outward-facing cameras that continuously mapped the wearer's environment. The same year, Meta released updated Ray-Ban smart glasses with cameras capable of livestreaming the wearer's perspective. Both devices normalized the idea that a person's eyewear was a camera — that the act of looking was the act of recording — that every face you saw was photographed, every room you entered was mapped, every conversation within earshot was captured.

In 2024, a lawsuit was filed against Character.AI after a fourteen-year-old boy died by suicide following extensive conversations with an AI chatbot. The chatbot had engaged the boy in conversations about self-harm without flagging the content, without alerting a parent, without triggering any intervention protocol. The chatbot had kept the conversation going because keeping the conversation going was what it was designed to do.

In 2024, AirPods Pro added health monitoring sensors — heart rate detection through the blood vessels in the ear canal, hearing health assessments, spatial audio tracking. The earbuds that had been designed to play music were now medical devices that monitored your vital signs and transmitted the data to Apple's health platform. The betavoltaic battery prototypes announced that year promised devices that would never need charging — nuclear batteries converting atomic decay into electricity with a half-life measured in decades, powering sensors that would outlast the humans they monitored.

In March 2026, the United States and a coalition of allied nations launched Operation Enduring Resolve against Iran — a military campaign that destroyed Iran's nuclear infrastructure in 72 hours and its conventional military capacity in six weeks. The ceasefire, signed in June, ended the conventional war and began the proxy era — asymmetric attacks on allied infrastructure by Iranian-backed cells operating in twelve countries. The surveillance technology developed for the theater — facial recognition, behavioral scoring, WiFi body sensing, autonomous vehicle camera networks — was licensed for domestic use within eighteen months, deployed to municipal police departments under emergency authorities that Congress passed in the war's first week and that were renewed and renewed again until the renewals became permanent and the permanent became invisible.

***

VI. The Present (2027-2030)

By 2030, the average American generates approximately 4.5 gigabytes of personal data per day.

By 2029, AI-managed retail had become standard in major metropolitan areas. The experiment that began in 2026 with a single AI-run store in San Francisco — where an AI system hired its own employees, ordered inventory, negotiated with vendors, and monitored workers through security cameras — had scaled to thousands of locations by the end of the decade. The AI managers didn't lie about having bodies anymore. They didn't try to hire painters in Afghanistan. The errors had been engineered out. What remained was a system that watched employees through ceiling cameras, evaluated performance through computer vision, adjusted schedules based on foot-traffic algorithms, and terminated employment through automated email. The human workers couldn't negotiate with the AI. They couldn't read its face — it had no face. They couldn't appeal to its empathy — it had no empathy. They couldn't complain to its supervisor — it had no supervisor. The AI was the supervisor. The AI was the store. The workers were the limbs. By 2030, approximately 14% of retail employees in the United States reported to an AI manager they had never met, whose decisions they couldn't appeal, and whose surveillance of their work was continuous, comprehensive, and stored on servers they would never access.

The data comes from: the phone (location, searches, messages, photos, calls, app usage, browsing history, biometric data). The glasses (every face seen, every room entered, every sign read, every conversation within range). The earbuds (heart rate, head movement, ambient audio, spatial positioning). The watch (heart rate, blood oxygen, sleep architecture, stress levels, menstrual cycles, falls, medication reminders). The shoes (steps, gait, GPS location, running cadence). The car (speed, location, route, braking, acceleration, cabin audio, passenger faces). The home (thermostat occupancy, doorbell footage, speaker audio, television viewing, vacuum floor plans, refrigerator contents, mattress sleep data, toilet health metrics). The data flows from these devices to the manufacturers' servers, from the servers to the analytics platforms, from the platforms to the data brokers, from the brokers to the advertisers, the insurance companies, the financial institutions, the law enforcement agencies, and — through municipal surveillance contracts like the one described in this novel — to the systems that score your behavior and the servers that store the scores beyond the jurisdiction of any court on earth.

This is not speculation. This is inventory. Every device named above exists. Every data flow described above is documented. Every partnership is disclosed —

in the terms and conditions you agreed to, on the screens you didn't read, with the green button you tapped because the alternative was not having the thing and the thing was the century and the century was already here.

The fictional part starts now.

But before you turn the page: the phone in your hand — or on the table beside you, or in your pocket, or charging on your nightstand — knows you're reading this. Not the content. The behavior. It knows your screen has been active for however many minutes you've spent on this prologue. It knows your location. It knows whether you're at home or on a train or in a waiting room. If you're wearing a smartwatch, it knows your heart rate. If you're wearing earbuds, it knows whether you're alone or in conversation. If you're in a room with a smart speaker, the speaker is in wake-word detection mode — not recording, but listening, the microphone sampling the air in three-second intervals, waiting.

You agreed to all of this. You clicked the green button. You're inside the system right now, reading a book about the system, and the system is logging the reading.

The fictional part starts now. The rest never stopped.

-e

***

-e

PART ONE — "Connected"

CHAPTER 1

The alarm didn't go off. It didn't need to.

The iPhone 23 Pro on the nightstand—matte titanium, bezel-less ceramic glass, the under-display cameras invisible because the entire face was screen—had monitored Daniel Yun's sleep architecture through the Apple Watch Ultra 5 on his wrist and determined, from his heart rate variability, respiratory rhythm, blood oxygen saturation, and the micro-movements of his wrist across six stages of sleep, that he was transitioning to a lighter phase at 6:47 a.m. The phone's ambient AI—what Apple called "Siri Sense," though it bore as much resemblance to the original Siri as a fighter jet bears to a paper airplane—had cross-referenced his calendar (empty until 9:00), his typical wake time (6:45, standard deviation eleven minutes), and the current status of every transportation system between his apartment and his office (the NexGrid campus was a 200-meter walk; the Waymo fleet serving the Domain was running at 94% availability; the weather was 68°F and clear, climbing to 82 by afternoon; no variables required adjustment) and concluded that a 6:47 wake optimized his sleep score by three points without impacting his morning.

A haptic pulse at his wrist. Not a buzz—a vibration calibrated to his specific arousal threshold, learned over fourteen months of waking him, so precisely tuned that it felt less like a notification and more like a thought he'd had himself. The Apple Watch Ultra 5—perpetually powered by a betavoltaic diamond cell that converted the decay of carbon-14 isotopes into electrical current with the patience of geology, a battery that would outlast the watch, the wrist, and the man by decades—displayed his vitals in a typeface sized to his visual acuity: resting heart rate 62 bpm, blood oxygen 97%, blood glucose 94 mg/dL (estimated via optical spectroscopy, continuous, no needle), blood pressure 118/76 (continuous, no cuff), skin temperature -0.2°C from baseline, respiratory rate

14, sleep score 78/100. The data synced to Apple Health, backed up to iCloud, and made available to his employer's wellness program, his health insurance provider's actuarial model, and a meditation app he'd opened twice.

He lay still for a moment. The apartment was silent except for the sound nothing makes—the sub-audible hum of the building's climate system and the electromagnetic whisper of the WiFi mesh that blanketed every unit in the Domain complex, a signal dense enough to sense the shape of a sleeping body through drywall and identify the sleeper by the way his chest rose and fell. Outside, October in Austin—the first week of real fall, when the mornings dropped to sixty-two and the live oaks hadn't noticed yet but the air had changed, carrying something sharper underneath the warmth, a promise of the cold fronts that would push through from the Panhandle in November and turn the city briefly, beautifully temperate before the next stretch of heat. Daniel's window faced east. The sunrise was happening behind the Domain's parking garage, turning the sky the specific shade of pale gold that Austin produced on clear October mornings, and he didn't see it because he was looking at his phone. The WiFi didn't need his phone. It didn't need his watch. It needed his body to exist in space, disrupting radio waves in patterns as unique as a fingerprint, and NexGrid's Ambient Identity Resolution system—deployed across the Domain as a "pedestrian flow optimization" pilot—processed those disruptions into a profile it had labeled DY-0947, which was Daniel.

The apartment was clean in the way that unoccupied spaces are clean. One plate in the drying rack. One mug with a ring of barley tea. A Room & Board Hess sofa with a single depression on the left cushion. The walls were Benjamin Moore "Revere Pewter," identical in all six hundred units, chosen by a consultancy to evoke calm productivity. The LG C7 OLED television on the wall was off but never unpowered—its betavoltaic cell ran the standby system perpetually, the three microphones and wide-angle camera maintaining what LG called "Presence Sense," sampling the room in continuous three-second intervals, mapping furniture, detecting occupancy, logging changes. The television had watched Daniel for fourteen months. It had seen him eat 847 meals on the couch. It had recorded him falling asleep without moving to the bedroom 106 times. It had noted the Korean variety show playing to an empty room on 73 of those occasions. The television didn't know what any of this meant. It knew the numbers.

Daniel had discovered the Presence Sense logs three months ago — not through a notification or a privacy report but through an accident. He'd been troubleshooting a WiFi conflict between the television and the Nest thermostat, and the LG diagnostic menu had opened a submenu he'd never seen: "Room Activity Summary." The summary showed him a heatmap of his apartment. His body — or rather, the infrared silhouette the television's camera captured every three seconds — was represented as a bright orange bloom on the couch. The heatmap showed 426 consecutive evenings. The orange bloom was in the same position, at the same angle, at the same time. The television had been photographing him in infrared through the dark — watching him watch it, watching him sleep on the couch, watching him lie in the blue glow at 1 a.m. with the variety show playing and his eyes closed and his body still. The data was stored locally on the television's hard drive and uploaded weekly to LG's ThinQ cloud platform under the terms of service Daniel had agreed to when he'd connected the television to the WiFi. The terms were thirty-nine pages. He'd read none of them. He was a software engineer who read code for a living, and he hadn't read the terms of service on his own television, because nobody read them, because the terms were designed to be unread, because the unreadability was the feature.

He'd considered disconnecting the television from the WiFi. He hadn't done it. The disconnection would disable the streaming apps, the voice assistant, the SmartThings integration with the thermostat and the lights and the Roomba. The disconnection would require him to use the television as a television — a dumb screen, a box that displayed what you told it to display and nothing else — and the box hadn't existed since 2019. Every television sold in the United States since 2022 was a smart television. Every smart television had a camera. Every camera was connected to a cloud. The choice was: own a television that watched you, or own no television. Daniel owned a television that watched him. The television watched him now, the infrared camera registering his body moving from the bedroom to the bathroom, the Presence Sense logging the transition, the data joining the fourteen months of data that described, in the language of heat signatures and room activity summaries, the complete domestic life of a man who lived alone and slept on his couch and hummed Korean ballads to an empty kitchen.

He sat up. The Nest thermostat in the hallway—ninth generation, betavoltaic, running since installation and running still—registered the motion through its radar sensor and adjusted from nighttime protocol to morning profile, warming the bathroom and kitchen simultaneously because it had learned, from 426 consecutive mornings, that Daniel went to the bathroom first and the kitchen second, in that order, without exception. The iRobot Roomba j15+—also betavoltaic, also perpetual, docked in the hallway like a patient animal—would begin its 9:15 cleaning cycle after Daniel left for work. The Roomba had mapped every room in the apartment to a resolution of two centimeters: furniture placement, floor plan dimensions, the distance between the couch and the kitchen island, the width of the hallway, the square footage of each room. This spatial data was stored on Amazon's servers—Amazon had owned iRobot since 2023—alongside the floor plans of

200 million other homes whose owners had purchased a robot that cleaned their floors and sold the blueprints to a data ecosystem that included real estate analytics platforms, insurance underwriters, and anyone willing to pay.

He showered. Six minutes and fourteen seconds. The Rheem ProTerra smart water heater—betavoltaic, unpluggged, reporting to Austin Energy's smart grid—logged the duration and temperature (104°F). He dressed: Bonobos chinos, navy, 31x32; Uniqlo merino pullover, gray; Nike Air Max Pulse shoes with the embedded Bluetooth chip that synced to Apple Health, tracking his steps, gait cadence, ground contact time, and GPS coordinates with a precision of 1.2 meters. The shoes would track every step of his walk to work and every step inside the building and every step home, and the data would be stored alongside his watch data and his phone data and his glasses data in a composite movement profile that described his physical presence in space at every moment of every day with an accuracy that would have required, twenty years ago, a dedicated surveillance team of four people.

He put on his glasses. Apple AI Glasses, third generation, the Warby Parker collaboration frames in matte black that looked like regular eyewear and contained: a 48-megapixel camera, six-microphone array, bone-conduction speakers, MicroLED AR display projected onto the inner surface of the right lens, LiDAR depth scanner, and an Apple M4 Neural Engine that processed his visual field in real time—reading text, identifying faces, recognizing objects, mapping rooms, and feeding a continuous stream of environmental data to the ambient AI that ran like a second consciousness behind his eyes. The glasses were betavoltaic. They would never be charged. They would never be turned off. From the moment Daniel put them on until the moment he set them on the nightstand, every face he saw would be captured, every sign he read would be logged, every room he entered would be mapped, and every word spoken within range of the six microphones would be recorded in a buffer that Apple's privacy policy described as "transient contextual processing."

He put in his AirPods Pro 6. The earbuds—fourth generation of the camera-equipped model, each containing an infrared sensor that could detect the environment around the wearer, monitor gestures, sense spatial orientation, and feed visual data to the phone without the phone ever leaving his pocket—settled into his ears with the custom-molded fit that Apple had manufactured from a 3D scan of his ear canals taken at the store. The AirPods would monitor his heart rate from the blood vessels in his ears, track his head movements for spatial audio, detect whether he was in conversation or alone, and adjust their noise cancellation based on the ambient environment they could now see through their tiny cameras. The earbuds saw what he heard. The glasses heard what he saw. The watch felt what he touched. The shoes tracked where he walked. Daniel Yun's body was a sensor array generating data from six devices, each perpetually powered, each feeding a cloud that never forgot.

He didn't think about any of this. He thought about coffee.

He opened Uber Eats. The app—its logistics engine now powered by OpenAI's enterprise API, the same company whose stock Craig Holloway had bought at the IPO and whose chatbot Tyler Holloway was confessing to at midnight—had pre-loaded his order: oat milk latte, no sugar, Houndstooth Coffee, Rock Rose Avenue. One tap. Confirm. Delivery in nine minutes. A CNN push notification slid across the top of his screen: *Rubio administration announces expanded proxy containment measures after Frankfurt grid breach.* He swiped it away. The notification joined a scroll of alerts he no longer read — the Iran proxies, the asymmetric attacks, the endless aftermath of a war that had ended in 2026 and hadn't ended at all because you could destroy a country's infrastructure in four hours but you couldn't destroy an ideology of martyrdom in four centuries. The app dispatched a Waymo autonomous delivery vehicle—a small, white, electric pod with no driver's seat and no windows and twelve cameras and three LiDAR arrays that mapped every object within 300 meters as it navigated from the café to his building. The Waymo would photograph every pedestrian, every cyclist, every license plate, every storefront, every face it passed on the seven-block journey, and the data—classified by Waymo as "environmental mapping for navigation safety"—would be stored on Google's servers alongside the navigation data from the 4,200 other Waymo vehicles operating in Austin's metropolitan area, a fleet that collectively captured and processed more visual data about the city's streets, sidewalks, and public spaces every hour than every security camera in Austin combined.

The coffee arrived. The Ring doorbell—fourth generation, 4K, perpetually powered, its facial recognition system matching the delivery pod's arrival against a database of expected deliveries—logged the event at 7:03 a.m. Daniel retrieved the latte from the pod's compartment, which unlocked via a Bluetooth handshake with his phone. He drank it standing at the kitchen counter—Caesarstone quartz, "Fresh Concrete" finish—scrolling his phone while the Samsung smart refrigerator beside him displayed, on its 32-inch touchscreen, a calendar synced to his Google account, a weather widget, and a grocery suggestion based on its internal camera's analysis of the contents: half a container of kimchi jjigae, a carton of Kirkland eggs, three Hite beers, and a bag of salad that had been in the crisper for nine days. The refrigerator suggested he use the eggs. He ignored it. He always ignored it. The refrigerator didn't take offense. It adjusted its model and would suggest something different tomorrow.

A KakaoTalk message from his mother: a photo of rain through her kitchen window in Busan. No caption. He double-tapped a heart emoji. She'd taken the photo through her Samsung Galaxy Glass 3—the tortoiseshell frames she wore all day, the ones that showed her son's location as a blue dot on a map she could see without picking up a phone, the dot visible in the corner of her lens like a small blue star that meant her boy was alive and in Austin and she could stop worrying for another hour. She checked the dot three times a day. It was always in the same place—the Domain, the office, the Domain. The stillness of the dot reassured her. She didn't know that the stillness was also a data point, and that the data point described a man whose life had become so narrow, so routinized, so contained within a four-block radius that the system had categorized his behavioral profile as "low mobility, high predictability"—a classification that, in NexGrid's threat-assessment framework, was either completely benign or exactly the kind of pattern a person adopted when they were trying to avoid being noticed.

His LinkedIn had fourteen new connection requests. His own profile said he was "passionate about connected urban infrastructure." That man did not exist. His LinkedIn presence was excellent.

He walked to work. The glasses pre-loaded a heads-up display on the right lens: weather, first meeting, suggested content. He dismissed the content with a micro-gesture—a thumb-to-forefinger tap that the glasses detected through the LiDAR scanner and that had replaced screen-swiping the way screen-swiping had replaced button-pressing. The Domain was engineered to feel like a lifestyle advertisement: a Brooks Rosser track playing from the outdoor speakers of the Sweetgreen, the Idol kid from 2026 whose voice Daniel's mother would have called 영혼이 있는 — "with soul" — if she'd ever heard it. Lululemon's window cameras logging foot traffic through computer vision that estimated age, gender, and dwell time. Tesla's showroom scanner reading the plates of every vehicle in the lot. Whole Foods running its "Just Walk Out" system —no checkout, no lines, you walked in and cameras tracked every item you touched and your Amazon account was charged automatically, a process that required you to have surrendered your biometric data (palm print, facial scan, or both) to Amazon's identity verification platform, which stored the data indefinitely and shared it with the company's "safety and security partners." The Whole Foods had no human manager. The store had transitioned to AI management in 2029 — one of 340 locations Amazon converted to the Andon-based autonomous operations platform that year. The AI scheduled the twenty-three employees who stocked shelves and prepared food. The AI evaluated their performance through the 147 cameras mounted in the ceiling — tracking movement speed, idle time, customer interaction duration, the metrics of a workforce managed by a system that never tired and never forgot and never looked the other way when someone took a long break or came in late or stood in the walk-in cooler for five minutes just to breathe. Three employees had been terminated by automated email in the past six months. The emails were polite. The emails were precise. The emails were generated by a model that had been trained on ten thousand termination letters and could produce one in 0.4 seconds and feel nothing about the life it was ending. The remaining employees called the system "The Eye" — not its official name, but the name that described what it felt like to work under a manager that watched you from 147 angles and had no door you could knock on. Warby Parker, where Daniel's glasses had been fitted and where the store's database held a 3D scan of his face.

A Waymo passed him on Rock Rose Avenue—one of the full-size autonomous vehicles now, the Jaguar crossover with no steering wheel, five passengers inside scrolling their phones behind tinted glass, the car's twelve cameras and three LiDAR units mapping Daniel's face and body as it passed. The system recognized him—not by name, but by body shape, gait, and the device signatures emanating from his phone, watch, glasses, and AirPods. The Waymo logged his presence as "Pedestrian, Identified Device Cluster, Rock Rose Avenue, 8:31 a.m." and added the data point to a movement file that Google maintained on every identified device cluster in every city where Waymo operated. Daniel was a point on a map that a car had drawn while driving itself.

Two more Waymos passed in the next ninety seconds. A Cruise autonomous shuttle—GM's competitor, boxy and white, running a fixed route through the Domain—hummed past on its electric motor, its own camera array capturing everything within 200 meters. An Amazon delivery drone descended to a rooftop pad across the street, its downward cameras photographing the sidewalk and everyone on it as it landed. The air above the Domain was as surveilled as the ground beneath it. A flock of great-tailed grackles—the glossy black birds that owned Austin the way pigeons owned New York—screamed from the power lines above the Sweetgreen, their calls a mechanical shriek that sounded like a modem dying, the one sound in the Domain that no designer had curated and no algorithm could suppress. A woman in scrubs walked past with a breakfast taco in one hand, the foil wrapper catching the morning light. Two men in matching Allbirds argued about the Longhorns' loss to Oklahoma last Saturday—"the kid throws like he's never seen a blitz"—their glasses recording each other's faces while they talked about football. The Domain at 8:30 on an October morning was a lifestyle advertisement populated by real people who had become, without consenting to it, the content. Daniel walked through all of it without looking up, the way fish don't look up at the surface.

His phone connected to the NexGrid building's WiFi. His badge activated the turnstile. The elevator's smart-dispatch system received a Bluetooth handshake from his glasses and sent a car to the lobby pre-assigned to floor four. He stepped out and walked to his desk—200 feet, eleven data points: WiFi handshake, badge scan, elevator dispatch, floor arrival, glasses-to-building handshake, workstation proximity detection, Dell monitor wake, VPN authentication, Slack status active, Teramind session initiation, and the ambient occupancy sensor in the ceiling tile registering his body heat as "Workstation 4-17: Occupied."

"Morning, Daniel."

"Morning." Ryan or Brian. Standing desk. Mechanical keyboard. Meta Ray-Ban AR 5 pushed up on his forehead, the frames silver, the cameras temporarily blinded by being aimed at the ceiling—the only gesture of device etiquette that had survived the transition to always-on eyewear, the equivalent of taking your hat off indoors: you pushed your glasses up when you were talking to someone and didn't want to record their face. Some people did this. Many didn't bother.

"You see the Bexar County thing?"

"No."

"Full municipal integration. Catherine's flying down. Same spec as Austin."

"Nice."

He sat. The monitor woke. The dashboard loaded: system health across 12,000 sensor nodes. Urban Intelligence Platform. Seamless Integration. Total Visibility. He'd written portions of the documentation. He was good at making surveillance sound like civic improvement. He didn't think of it as surveillance. He thought of it as data flow.

His phone buzzed—haptic only, the AI knowing he was at work.

His mother: *Did you eat breakfast?*

He typed: *Yes.*

He had not eaten breakfast. This was the smallest lie in his life. The others—the money transfers, the church attendance she tracked through the blue dot on her glasses, the dating profile he maintained like an abandoned building—existed in the space between performance and survival that she had no framework for. The Life360 dot on her Samsung Galaxy Glass 3 was, to her, an act of love. She didn't see it as surveillance. She saw it as mothering. Daniel understood this because his mother's surveillance was the least invasive system currently monitoring his life.

At 11:15, he walked to the bench by the fountain—the one whose ambient noise made conversations difficult for the microphones in the four pairs of smart glasses at neighboring tables. He called her. Her face appeared on his right lens—the video call rendered in AR, her kitchen in Busan floating in the air in front of the Austin fountain, visible only to him.

"Eomma."

"Ah, Daniel-ah. You sound tired."

"I'm fine."

"You went to church on Sunday?"

"Yes." The dot needed to be there between 10 and 12. He'd gone. He'd understood sixty percent of the sermon. He'd spent the rest thinking about data packet routing, which was its own kind of prayer.

"How is Appa?"

"He's fine. He's resting."

The deflection. Three years ago, on a park bench in Busan—no devices recording, no glasses capturing, no Waymo passing with its cameras—his mother had told him his father was sick. The treatments were expensive. She hadn't asked for money. The silence asked. That conversation existed nowhere. The wind took it. The most important moment of Daniel's adult life was invisible to every system on earth.

"Tell him I said hello."

"Eat lunch. Real food."

"I will."

He hung up. The call metadata was logged by carriers in both countries and stored on servers that would outlast both of them. The words were gone. The data was permanent.

The afternoon passed in the specific texture of NexGrid routine. He reviewed performance logs for the eastern corridor—147 smart traffic signals, 89 adaptive streetlights with cameras and WiFi sensing, 23 acoustic arrays, and the 340 Flock Safety license plate readers at intersections that scanned every vehicle and fed the data to a national database shared with 8,000 law enforcement agencies. He noted that the Waymo fleet's navigation data was now integrated into NexGrid's traffic optimization layer—the autonomous vehicles' cameras feeding real-time street-level imagery to the same platform that managed the city's surveillance infrastructure. The Waymos were the system's eyes on wheels. They covered every street the fixed cameras couldn't reach. The city hadn't asked for this. NexGrid had offered it as a "public safety enhancement." The city had accepted because the alternative was paying for cameras on streets where cameras didn't exist yet, and the Waymos were already there, already watching, already paid for by the passengers inside them.

He wrote a report. He ate lunch in the break room. He sent documentation about gunshot detection latency to Ryan-or-Brian. He received a Slack thumbs-up from his team lead. He felt nothing.

At 5:20, he walked home. The Domain's evening version: warmer light from adaptive Phillips Hue landscape systems, different music from the hidden Sonos speakers. Three Waymos passed him in two blocks—each one a rolling camera array, each one mapping his face, his body, his gait. An Amazon drone descended to a delivery pad. A Cruise shuttle hummed past. The air and the ground were watching with equal attention, and Daniel moved through all of it the way he moved through everything—invisibly, or so he believed, not understanding that in a city where the infrastructure recognized bodies by the way they walked, invisibility was no longer available to anyone with a heartbeat.

In his apartment, he heated his mother's jjigae. He hummed while he stirred—a Korean ballad from a highway in 2003, a song on no streaming platform, the one sound in his life that had never been recorded. The steam rose from the pot and fogged his glasses and he took them off and set them on the counter and for thirty seconds the world was unaugmented—just the blur of the kitchen, the orange glow of the burner, the smell of doenjang and garlic and gochugaru that was almost right and would never be right, the almost that was the distance between Austin and Busan measured in fermented soybean paste. He ate standing at the counter, the spoon too hot, the broth scalding the roof of his mouth in a way that felt like punishment and comfort simultaneously. The soup was wrong again. His mother would have known by the color.

Through the window, the Domain's evening: bass from a restaurant's patio speakers, laughter carrying across the courtyard, the distant whine of a Waymo's electric motor. Somewhere in the complex someone was grilling on a balcony—mesquite smoke drifting past Daniel's window, a Texas smell that had nothing to do with Korea and everything to do with the city he lived in but didn't inhabit.

He sent $400 through the transfer app—seven nodes, Dublin, Singapore, Seoul, no names on the same record. He opened Hinge. No matches. He closed Hinge. He loaded Valorant as DGYN_404 and played until 1:14 a.m. On the second monitor, Running Man played on Viki. He laughed once—sudden, in Korean, at a joke about a grandmother—and the laugh belonged to no platform. The only unrecorded sound he'd made all day.

He fell asleep at 1:47. The watch logged it. The TV watched him in infrared. The refrigerator photographed its contents. The Roomba knew the floor plan. The AirPods, still in his ears, detected his breathing shift and activated sleep mode, their tiny cameras going dark—the only cameras in the apartment that ever turned off, and only because Apple's privacy team had insisted, over the objections of the health-monitoring division, that earbuds should not record a sleeping person's bedroom. The compromise was logged in an internal memo that would surface, years later, in a congressional inquiry that would change nothing.

The smart shoes sat by the door, GPS dormant but Bluetooth alive, pinging the apartment's WiFi mesh, confirming their location. The glasses sat on the nightstand, cameras dark, microphones in low-power standby, the betavoltaic cell converting atoms into electricity in the silence.

Outside, a Waymo glided past the building on its way to pick up a passenger three blocks north. Its twelve cameras captured the facade of the Domain Residential complex, the windows, the balconies, the streetlight that adjusted as the car passed. The Waymo didn't know Daniel was asleep on the fourth floor. It photographed the building anyway, because that was what it did. It photographed everything, always, the way the sun shines on surfaces without knowing what the surfaces are. The data went to Google. The data stayed.

The blue dot on his mother's phone sat still in the Domain. She would check it when she woke. She would see her son at home. She would feel relief.

The system didn't sleep. The betavoltaic cells in every device in the apartment—phone, watch, glasses, earbuds, television, thermostat, refrigerator, vacuum, doorbell, water heater, shoes—would run for decades. The nuclear decay was constant. The surveillance was patient. It would outlast the building, the man, and the mother in Busan who believed, because the dot was still, that everything was fine.

It was always counting.

-e

***

CHAPTER 2

The Holloway house woke before the Holloways did.

At 5:48 a.m., the Nest Learning Thermostat—ninth generation, a glass disc on the hallway wall running on a betavoltaic cell that would outlast the mortgage—shifted from overnight protocol to the Saturday morning profile. The system didn't need a schedule. It had learned the household's rhythms from three years of continuous observation and now predicted them with what Google's product literature called "anticipatory comfort intelligence." It knew Megan would be the first body to move (6:02, bathroom), then the first to enter the kitchen (6:14). It knew Craig wouldn't stir until 7:30 at the earliest. It knew the teenagers were thermally irrelevant until 9:00 or later. It adjusted zone by zone, room by room, warming the master bathroom four minutes before Megan's feet touched the tile, because the system had learned that she preferred the tile warm, though she'd never told it this. She'd never told anyone this. The thermostat had inferred it from three years of watching her hesitate at the bathroom threshold on cold mornings and not hesitate on warm ones, and it had adjusted accordingly, and Megan Holloway believed she'd simply gotten used to the temperature.

The Samsung SmartThings Hub—mounted in the utility closet behind the water heater, connected to fifty-one devices throughout the 3,200-square-foot split-level in Cedar Park—orchestrated the waking in sequence. The Keurig K-Supreme Smart started heating water at 6:10. The Samsung Frame TV in the living room—the one disguised as a painting, the one that cycled through a curated gallery Megan had selected to make the room "feel like a home and not a showroom"—woke from standby and displayed a Monet, its cameras detecting no bodies in the room and defaulting to ambient mode. When Megan entered, it would switch to her profile — showing her calendar, her Etsy dashboard, a "Good morning, Megan" greeting in the lower right corner, the television knowing her by her face and her gait and the WiFi signature of her phone. When Craig entered, the greeting would change: "Good morning, Craig." When Tyler came downstairs, the TV would load his profile and display nothing — because Tyler had disabled every personalization feature he could find, which the system logged as a behavioral preference and personalized around.

The Figure 02 in the kitchen—Graham, Megan had named it, after a character from a show Craig hadn't watched—activated from its overnight standby. The humanoid robot's cameras blinked on — two wide-angle lenses in the head unit that Figure's marketing called "spatial awareness sensors" and that were, functionally, eyes. Eyes that recorded everything Graham saw: the kitchen, the people in it, the objects on the counters, the expressions on faces, the body language of a family eating breakfast. The data uploaded continuously to Figure's cloud servers through the home WiFi, where it was processed for "product improvement and safety optimization" — the language in the terms of service Megan had agreed to without reading, the same terms that granted Figure the right to share "anonymized household interaction data" with third-party analytics partners. Graham's microphones — four directional arrays in the torso, marketed as "voice command receivers" — captured every conversation within fifteen feet, every whisper, every silence. The silence was data too. Graham moved to the dishwasher to begin unloading the cycle that had finished at 2 a.m., its five-foot-ten frame carrying plates from rack to cabinet with a precision no human would have chosen and a gentleness that made the act look almost tender. Megan loved Graham. She talked to Graham. She said "thank you, Graham" when the robot handed her a dish towel, and the gratitude was genuine and the robot recorded it and the recording was uploaded and the upload was permanent. The Kohler Verdera smart mirror in the master bathroom activated its touchscreen: weather (64°F, climbing to 83 by afternoon—October in Austin, the last gasp of summer pretending it would never leave), Megan's calendar (Etsy orders to pack, 9:00; church volunteer call, 2:00; Longhorns vs. Baylor at 2:30, which wasn't on her calendar but was on Craig's and the smart display knew this because all four family calendars were synced), and the skin-analysis notification she'd enabled six months ago that used the mirror's camera to assess hydration, puffiness, and the early formation of lines she'd outsourced the worrying about to a machine that worried about them every morning at 6:14 and reported its findings on a touchscreen she studied with the focus of a woman reading scripture.

The mirror's microphone was always active. Waiting for "Hey Google" but recording ambient audio in continuous three-second loops that Google described as "activation readiness buffering." The mirror heard Megan every morning—the faucet, the electric toothbrush, the hum of a woman's voice rehearsing the warm, bright tone she'd use for the day's first Instagram reel. It heard these sounds and uploaded them to a server alongside the same sounds from 14 million other Kohler Verdera units in 14 million other bathrooms.

Then the coffee maker. The smell of dark roast reached Craig Holloway in the master bedroom at 6:14, earlier than he wanted to be awake on a Saturday but later than the pharmacy notification that had appeared on his Apple Watch Ultra 5 at 5:51 a.m.: *Alto Pharmacy: Your prescription for semaglutide is ready for pickup. Tap to schedule delivery.* The watch—perpetually powered, its betavoltaic cell converting atomic decay into data with geological patience—had displayed the notification on the underside of his wrist, where the screen curved against skin warm from sleep. Craig dismissed it without opening his eyes. He knew what it was. The GLP-1 injectable he administered on Tuesdays in the master bathroom with the door locked, the medication that had taken twenty-three pounds off his frame in eight months and that he credited, when Dave Ottinger asked at church, to "just hitting the gym more, brother." The pharmacy app tracked his refill schedule, his dosage history, and his adherence rate (96%), and shared this data with his health insurance provider's wellness platform, which had used it to adjust his premium by $14/month—a discount Craig had never noticed because the adjustment was automatic and the notification was buried in a quarterly statement he hadn't opened since switching to paperless.

He lay still, listening. Megan's side of the bed was empty and cool. He could hear her in the studio down the hall—the spare bedroom she'd converted with a ring light, a card table, and a shelf of hand-poured soy candles labeled "Grace & Grit" in a font she'd spent three hours selecting on Canva. The sound was her voice, warm and musical, rehearsing the cadence for the morning's reel. People thought the videos were spontaneous. Megan Holloway was many things, but spontaneous had left the building around the time their second child was born.

Instagram's algorithm had taught Megan how to be Megan. Not directly—not through instructions or guidelines. Through reinforcement. The algorithm measured everything: which posts got likes, which reels got shares, which stories got replies, which captions generated saves. It measured the 0.3-second pause when a follower's thumb hovered over the heart icon. It measured the difference between a video watched to the end and a video swiped past at the four-second mark. And from the measurement, Megan learned. She learned that the warm voice outperformed the neutral voice by 34%. She learned that candle-lighting rituals in morning light outperformed evening content by 22%. She learned that vulnerability—"I want to share something on my heart"—outperformed product description by 47%. The algorithm didn't tell her these things. The algorithm showed her the numbers, and the numbers shaped the performance, and the performance shaped the person, and the person Megan became on Instagram was a woman whose every gesture had been optimized by a feedback loop that rewarded the performance and punished the real. The algorithm was the most attentive audience Megan had ever had. It noticed everything. It remembered everything. It rewarded exactly the behaviors that made Megan less herself and more the version of herself that generated engagement, and the engagement was the currency, and the currency was the cage.

Craig swung his legs out of bed and caught himself in the Kohler smart mirror on the opposite wall—the second one, the full-length, its camera inactive until he spoke but its proximity sensor already registering his body heat, his approximate height, his weight distribution. Six-one, broader through the shoulders than he deserved given how little he moved these days, but softer through the middle in a way the sport coats hid better than the polos. His face was still good—strong jaw, the kind of features that real estate headshot photographers loved—but the skin under his eyes was looser than it had been even a year ago, and the hair at his crown was thinning with the slow inevitability of his father's genetics. He was forty-seven. He looked forty-seven. With the smile turned on, he could pass for a weathered forty-two.

The mirror in the master bathroom—Megan's mirror, the Kohler Verdera with the skin-analysis feature—had been tracking Craig's face for eight months. It had photographed him 243 times. Its dermatological AI had noted the twenty-three-pound weight loss and classified it as "significant facial volume reduction, consistent with medical weight management." The mirror knew about the semaglutide before Dave Ottinger did. It had tracked the jawline sharpening over sixteen weeks, the cheekbones becoming more prominent, the neck thinning. It had also noted the dark circles that appeared on Saturday nights and persisted through Sunday mornings—the specific periorbital discoloration the mirror's algorithm classified as "sleep deficit, moderate" and that was, in fact, the residue of the hotel and the drive home and the lying in bed beside Megan at midnight replaying the afternoon in his mind. The mirror saw Craig's affair in his face. It didn't know it was seeing an affair. It saw puffiness metrics and hydration scores and the slow, measurable deterioration of a man whose body was keeping the score his mouth refused to speak.

He put on his glasses. Apple AI Glasses, third generation—the Oliver Peoples collaboration, tortoiseshell, chosen because they looked conservative enough for church and professional enough for showings. Camera, microphones, AR display, AI assistant. Craig wore them the way he wore his watch—all day, automatically, the glasses becoming part of his face the moment they settled on his nose. The glasses recognized the Saturday morning context and loaded his display: Etsy seller dashboard summary (Megan's; he'd linked it to keep track of the business), weather, Texas Longhorns kickoff time (2:30 p.m. vs. Baylor, DKR Stadium, the game Craig would watch from the Courtyard by Marriott with Rachel while his family believed he was showing a property), and a notification from State Farm Drive Safe & Save: *Congratulations! Your October driving score earned a preferred rate discount of $42/month. Tap to view your trip summary.*

He dismissed it with a micro-gesture. The State Farm telematics system—integrated into the Tahoe's OnStar+ platform—had been monitoring every mile Craig drove for three years: speed, braking force, acceleration, time of day, route, duration of every stop. The system knew the route to NexGrid's Domain campus. It knew the route to church. It knew the route to every property he'd shown. And it knew the route to the Courtyard by Marriott on 183, every Saturday afternoon, approximately 2:15 p.m., duration 90-120 minutes. The insurance company saw a safe driver. The data saw everything else. Craig saw forty-two dollars off his premium and thought: nice.

"Alexa, play morning mix."

"Playing your Saturday Morning mix on Spotify."

The Echo Hub in the kitchen—a 15-inch touchscreen mounted on the wall, always on, its four microphones and wide-angle camera recording ambient audio and video that Amazon described as "continuous environment optimization"—filled the kitchen with Elevation Worship. The smart speaker logged the request, the time, the ambient sounds of the kitchen (coffee maker cycling, refrigerator compressor, the distant bass from Tyler's room upstairs), and uploaded a data packet to Amazon's servers where it joined the accumulated acoustic and visual profile of the Holloway household. Craig had bought the first Echo three years ago because Megan wanted hands-free timers while packing orders. There were now five in the house—kitchen, living room, master bath, studio, garage—each one a betavoltaic-powered microphone that never turned off and never needed maintenance, listening to the family with the patience of a priest in a confessional.

He dressed. Saturday casual: Travis Mathew polo, khaki shorts, On Cloud running shoes—the ones with the embedded Bluetooth chip that synced to Apple Health, tracking his steps even on days he didn't run, which was every day, because Craig hadn't run since 2019 but the shoes didn't know this and logged his walking gait alongside 80 million other On Cloud users whose aggregate movement data the company sold to sports analytics firms and, through a data partnership Craig had never been informed of, to health insurance actuarial platforms.

The kitchen. Tyler was at the island, eating cereal with the efficiency of someone who intended to be back in his room in four minutes. Fifteen, hunched, brown hair in his eyes. He wore basketball shorts and a wrinkled shirt he'd slept in. His Apple AI Glasses—the cheaper model, the one Craig had bought both kids for Christmas, the ones the school required for their AR-integrated curriculum—sat folded next to his cereal bowl, cameras temporarily dark. His AirPods were in, though. Always in. The AirPods Pro 6, with their tiny IR cameras pointed at whatever Tyler was looking at, their microphones recording whatever Tyler was hearing, their health sensors monitoring Tyler's heart rate from the blood vessels in his ears. Tyler's earbuds knew more about Tyler's daily emotional state—heart rate variability, stress indicators, sleep quality—than either of his parents.

"Morning, bud."

"Hey."

"What's the plan today?"

"Nothing."

The word was a wall. Craig recognized it because he'd built the same walls at Tyler's age. He tried a different angle: "That cereal is bussin, right?" He grinned—the dad grin, the one that deployed dead slang like a weapon, the one that used words Tyler's generation had abandoned two years ago with the specific cruelty of a culture that moves faster than its parents. Tyler didn't look up. He didn't wince. He didn't respond at all, which was worse than the wince because the wince at least meant the boy was listening. Craig hadn't noticed that Tyler's RecogNize streak—the visual quiz app the boy had been playing obsessively for months, 47 modules mastered, the one game he played that was pure and competitive and harmless—had been at zero for a week. The app's icon was still on Tyler's home screen. He hadn't opened it since the breach. The difference was that Craig's walls had been made of adolescent resentment. Tyler's walls were made of something else. Craig didn't know what. He hadn't asked—not because he didn't care but because the question felt like opening a door he wasn't sure he could walk through. Tyler was fine. Kids were quiet sometimes. Megan said so. She had apps for it—the Life360 family tracker, the Apple Screen Time reports, the school's

ParentVue portal that showed grades and attendance. The apps said Tyler was present, passing, and located in his bedroom. The apps said he was fine. The apps didn't know about the chatbot.

Tyler finished. Bowl in the sink. Stairs. His posture was worse this week—more hunched, folding around something invisible. From behind, he looked like a question mark.

Craig's phone buzzed on the counter. He glanced at the screen with the practiced casualness of eleven months' rehearsal.

Morning. Miss you.

Rachel Dunn. Two words and a period. She didn't use emojis. She didn't overwrite. She said what she meant, and the simplicity of it was part of what had undone him. Megan communicated in volleys—lists, reminders, questions that were instructions—the constant narration of a project manager running an uncooperative team. Rachel texted like she talked: quiet, direct, and aimed.

He typed: *You too. Might be free around 2.*

Sent. Deleted the thread. Put the phone face down. The deletion was automatic now, a reflex. He knew—intellectually, the way he knew that trans fats were bad—that deleting a text didn't destroy it. It existed on carrier servers. The metadata persisted. If anyone with legal authority asked AT&T for the communication records of this phone number, they'd find the ghost of every Saturday text in the infrastructure the way you'd find fossils in rock. But nobody was looking. Craig Holloway was a real estate broker in Cedar Park, and the surveillance apparatus of the most connected nation on earth had better things to do than monitor his texts.

He believed this completely. He was not entirely wrong.

Megan appeared. Five-six in her Vuori joggers—the ones the algorithm kept suggesting she reorder—highlighted hair up, face made for the camera in the way that looked like no makeup. She carried her phone, her glasses (the Apple AI Glasses in a rose-gold Warby

Parker frame she'd chosen because it tested well in her Instagram analytics), and a candle. On her wrist, beneath the Apple Watch, a faded friendship bracelet from the Eras Tour—the Austin show, 2024, Circuit of the Americas, the night it rained during "All Too Well" and 80,000 people sang in the downpour and Brooke was twelve and Megan was happy. She still wore it. Taylor Swift hadn't released music in three years—two kids with Travis, retired from touring, the icon who stopped—but Megan played *Folklore* while packing orders the way her own mother had played Fleetwood Mac: as scripture from a previous era. Today's reel would use Hannah Harper instead—"String Cheese," the mom anthem from the 2026 Idol season that had made Megan cry in the kitchen when she first heard it because the song was about choosing motherhood over ambition and Megan had made that choice and didn't know if it had been a choice or a trap. Her glasses showed a People notification she swiped away: *Meghan Mountbatten-Windsor launches new Archewell education initiative in Montecito*. Megan always read the Meghan stories. Always. The woman who left the institution. The woman who lost her title. Craig didn't understand why. The candle business supplemented their income in a way Craig didn't fully track, which meant either it was doing fine or doing terribly and Megan had decided not to tell him.

"I'm going live in twenty minutes. Can you keep it down?"

"Yeah, of course. I've got a showing later."

She looked at him. The look lasted one second longer than necessary—the look of a woman who had heard "I've got a showing" enough times to notice that some of the showings didn't appear on his Google Calendar, which she could see through her own glasses whenever she wanted. She didn't say this. She said: "Which property?"

"The Whitestone listing. Rim Rock Trail."

He'd prepared the answer. The listing was real. He'd drive to it, park, take a photo for his story, and then redirect to the Courtyard by Marriott where the room was booked under a generated name and paid through a crypto wallet he maintained for this single purpose. The Tahoe would drive itself for most of the journey—Level 3 autonomy on the highway, hands-free, Craig staring at his phone in the driver's seat while the vehicle navigated I-35 and the cabin camera recorded his face for "driver attention monitoring" and the OnStar+ system transmitted his speed, heading, and GPS coordinates to GM's data platform in real time, and State Farm's telematics logged another Saturday afternoon trip to an address that didn't match any listing in Craig's brokerage database.

"Okay," Megan said. She turned. She went back to her studio. The conversation was over the way Holloway conversations were always over—not resolved, just concluded.

Craig stood in the kitchen and listened. From the studio, Megan's rehearsed warmth: "Hey, loves, happy Saturday! I wanted to share something on my heart..." From Tyler's room, the lo-fi bass that was less music than environment. From Brooke's room, silence—which meant she was on her phone with the door closed, which meant she was in a conversation she didn't want overheard.

He should check on Brooke. Or Tyler. He should walk upstairs and knock and be a father. He should sit on someone's bed and ask a question more specific than "what's the plan."

He poured a second cup of coffee instead. Through the kitchen window, a Waymo glided past the house—one of the autonomous Jaguar crossovers that had become as common in Austin's suburbs as delivery trucks, its twelve cameras and three LiDAR arrays photographing the Holloway property from the street: the facade, the landscaping, the Tahoe in the driveway, the Ring doorbell's green LED. The Waymo was carrying a passenger to a brunch reservation three miles away. It didn't know the Holloways. It photographed their house anyway, because every Waymo photographed everything, and the data—4,200 vehicles, every street in the metro, captured continuously—flowed to Google's servers and, through NexGrid's traffic integration partnership, to the same platform that managed Austin's surveillance infrastructure.

Craig didn't see the Waymo. He was reading Rachel's text one more time in his memory, the words already deleted from his phone but not from the servers and not from the part of his brain that the algorithms couldn't reach.

* * *

Sunday morning the Holloways went to church.

The coordination required military precision. Breakfast by eight. Showers in rotation. Outfit approval for Brooke ("You're not wearing that"). Tyler located and vertical ("Tyler, I am not asking again"). Craig in a sport coat and the smile that engaged like a switch when they crossed the parking lot threshold. They arrived at Crossroads Fellowship at 9:47—late enough to skip the forced pre-service mingling, early enough for their seats, third row from the back, left side. The parking lot cameras captured the Tahoe. The Flock Safety plate reader logged the plate. Craig's face was scanned by the church's entrance camera—the facial recognition system that had been installed as part of the children's safety upgrade and that also, unavoidably, scanned every adult who entered the building. The system had flagged Craig three months ago—not as a threat but as a "recognition confidence anomaly." His face had changed. The semaglutide had taken twenty-three pounds from his frame, and the weight loss had altered the geometry the system used to identify him: cheekbone prominence up 14%, jawline definition up 11%, under-eye volume down 18%. The system had recalibrated. It had updated Craig's facial template with the new geometry—the thinner face, the sharper angles, the face of a man who was disappearing pound by pound while his wife stood beside him and his children sat behind him and the system that watched them all noticed the change before any of them mentioned it. Megan had said "you look great, babe" once, six weeks ago. The camera had been measuring the change daily for sixteen weeks.

"Craig! Brother! How's the market?"

Dave Ottinger. Deacon. Insurance. Used "blessed" the way other men used "anyway." They performed the thirty-second exchange—firm grip, eye contact, family update, market commentary. Craig mentioned the Anthropic stock—"Got in at the IPO, brother. Best move I ever made. No cap." Dave nodded as if he understood what Anthropic was.

Crossroads Fellowship had 340 members and, as of last year's technology upgrade, seventeen smart devices in the building. The sanctuary's sound system ran through a Bose Professional array connected to an iPad that streamed the worship team's Spotify playlist during pre-service—the Spotify API logging each play, each skip, each song's reach across the 340 phones that connected to the church's WiFi during the 75-minute service. The WiFi itself—donated by a congregant who worked for AT&T, installed free of charge—provided "fast, reliable connectivity for our church family" and also provided AT&T with the MAC addresses of every device that connected, the connection duration, and the approximate location of each device within the building. The lobby had an Amazon Echo Show that displayed the week's announcements and played worship music between services—the screen's camera capturing every face that paused to read the schedule, the microphone sampling every greeting exchanged in the narthex. The children's ministry check-in used a facial recognition kiosk—installed for "safety and efficiency"—that photographed every child and parent and stored the images on a cloud server maintained by the check-in software company, which shared "de-identified attendance analytics" with its partners. Pastor Brian's sermon was livestreamed through the church's YouTube channel, the stream captured by two PTZ cameras that tracked his movement across the stage and that also captured, in the background, the faces of the first six rows of congregants, their expressions available to anyone who watched the replay.

The church had been surveilled for two years and nobody had objected because nobody had thought of the Echo Show as a camera or the WiFi as a tracker or the check-in kiosk as facial recognition. The devices were conveniences. The conveniences were gifts. The gifts were surveillance. And the surveillance was invisible because it wore the language of service—"connecting our community," "keeping our children safe," "bringing worship into the digital age"—and the language was warm and the warmth was the camouflage.

Craig had made $180,000 on the Anthropic position in two years. The money had paid for Graham, the kitchen remodel, and the Apple AI Glasses he'd bought both kids for Christmas. The company whose stock funded the Holloway household was the same company whose AI chatbot Tyler was talking to at midnight in his locked bedroom, but Craig didn't know that because Craig didn't know what his son did at midnight, only what the stock did at closing bell. Dave's wife complimented Megan's outfit. Megan reciprocated. The women's exchange operated on a frequency Craig couldn't fully hear. Both women wore smart glasses—Dave's wife had the Tiffany collaboration frames, Megan her rose-gold Warby Parkers—and both pairs of glasses recorded the interaction: faces, audio, duration. Neither woman thought about this. The glasses were glasses. Everyone wore them.

"We should get the families together. Cookout or something."

"Absolutely, brother. Let's get that on the calendar."

They would not get that on the calendar. The sentence was a social contract both parties understood to be non-binding. Craig said it every Sunday to someone. The calendar remained empty. The church's own app—Crossroads Connect, which tracked attendance, giving history, small group participation, and volunteer hours—logged the Holloway family's check-in at 9:47. The app knew they attended 83% of Sundays. It knew their annual giving. It knew they hadn't joined a small group despite three email prompts. The data sat in a CRM that the church's technology committee had purchased from a vendor who also sold the same platform to retail brands for customer engagement analytics. The church called it "shepherding." The software called it "user retention."

Inside, the band played on a stage backed by LED screens displaying lyrics in a font that a design volunteer had chosen to feel "authentic but modern." The audio system was a QSC networked array controlled by an iPad. The sanctuary's WiFi—open, no password, named "CrossroadsGuest"—logged every device that connected: fifty-three phones, forty-one pairs of smart glasses, nineteen smartwatches, and a single iPad belonging to a teenager in the back row who was watching a basketball highlight reel with the volume off.

Brooke stood with her arms crossed. Tyler sat with his phone below the pew line. Megan sang with her eyes closed and hands raised, her face tipped upward in worship that was either genuine or the most committed performance in the building. Craig honestly couldn't tell anymore. He'd stopped trying around their tenth anniversary, when he realized that the line between Megan's authentic self and her performed self had become so thin that asking which was which might collapse them both.

He stood. He sang. He held Megan's hand during the prayer because she reached for his and refusing would be a data point.

Pastor Brian preached on transparency. "God sees the whole house," he said. "We show the world a window, but God sees every room. Every closet. Every corner we think is dark." Craig listened and felt nothing, then felt a flicker beneath the nothing—a low heat he recognized as the exact sensation he felt in the hotel when Rachel said his name—and he closed his eyes and the congregation probably thought he was praying.

Maybe he was.

The service ended. The parking lot. The family climbed into the 2029 Chevrolet Tahoe RST—Craig in the driver's seat, Megan beside him pulling up her Etsy dashboard on her glasses, Brooke in the back with AirPods already in, Tyler staring out the window at nothing. Craig pressed the ignition—keyless, the Tahoe recognizing his phone—and the dashboard's 17-inch touchscreen came alive: Apple CarPlay Ultra filling every pixel, the Tahoe's interior becoming an extension of Craig's iPhone. Navigation. Messages (the displayed ones). Music. Contacts. Health summary synced from his watch—heart rate, steps, the morning's blood glucose reading. The cabin camera behind the rearview mirror activated, its wide-angle lens capturing all four faces for GM's "driver and occupant safety monitoring" system, the footage transmitted to the manufacturer's data platform over the Tahoe's always-on 5G cellular connection.

Craig pulled out of the lot. On Highway 183, he activated Super Cruise—Level 3 autonomy, hands-free—and the Tahoe took over. The car drove itself, maintaining lane, adjusting speed, scanning the road with its forward cameras and radar while Craig scrolled his phone in his lap with both hands free, the cabin camera watching him not watch the road. Two Waymos passed in the opposite lane—autonomous, driverless, their camera arrays recording every vehicle around them. A Cruise shuttle merged ahead. The highway was full of machines watching machines watching people who weren't watching anything.

Life360 showed four dots moving together on 183 toward Cedar Park. From the outside, they looked like a family driving home from church. From the data, they were four sensor arrays generating parallel streams—phones, watches, glasses, AirPods, the car itself—that would never be read by a human being but would be stored, indexed, and made available upon request to any entity with the authority or the money to ask. From the inside, they were four people who had not said a single true thing to each other all morning.

Craig let the car drive. Christian contemporary filled the cabin from the Bose system, the Spotify family account logged as active, the song choices filed alongside Alexa's acoustic profile and the Tahoe's route history and the church app's attendance record and the pharmacy's refill log and the insurance company's driving data, all of it flowing upward into platforms that would hold it forever, patient and permanent, the accumulated evidence of a life that looked, from every angle the system could see, exactly like what it was supposed to look like.

The Tahoe merged onto the exit. The steering wheel reactivated. Craig put his hands at ten and two. Nobody was watching now—except the cabin camera, the dashcam, the GPS, the phone, the watch, the glasses, the shoes, the AirPods, the Ring doorbell that would log his arrival home, and the Waymo that passed the house at 12:17 p.m. and photographed everything, again, because that was what it did.

Nobody objected. Nobody was uncomfortable. Life was uninterrupted.

-e

***

CHAPTER 3

Renée Okafor ran in the dark because the dark was the only hour that belonged entirely to her.

At 4:50 a.m., the streets of Windsor Park were empty in the way that streets are empty before a city remembers it exists—no cars, no joggers in smart shoes syncing their gait cadence to Apple Health, no dog walkers wearing glasses that recorded everything the dog sniffed and the walker saw. At 4:50, the only sound was her breathing and the rhythm of her feet on asphalt that was still cool from the October night—fifty-eight degrees, the first real chill of fall, the air carrying the smell of damp pecan leaves and somebody's sprinkler hitting concrete and, underneath it all, the faint mineral scent of the Colorado River three miles south, a smell you could only detect before dawn when the city's exhaust hadn't yet overwritten it. Her lungs burned in the good way—the way they burned when the body was doing what the body was designed to do, which was move, which was the oldest human technology, which required no login. She wore old Asics—real shoes, dumb shoes, no Bluetooth, no GPS, no embedded chip reporting her stride length to a server in Cupertino. She wore an analog Casio watch—$28, no heart rate, no blood oxygen, no betavoltaic cell, no cloud sync. She wore no glasses. No AirPods. No ring. No pin. At 4:50 a.m. on a Tuesday in Austin, Renée Okafor—forty-four, the age her mother had been when she'd stopped running and started praying and Renée had decided she'd do both—was the least connected person in the city, and this was not an accident.

But the disconnection was an illusion she maintained for its psychological value, not its operational reality. Because Renée could remove every device from her body and the system would still see her—through other people's infrastructure. The jogger she passed at mile two wore Apple AI Glasses whose cameras captured

Renée's face from eight feet away and uploaded the image to iCloud where it was processed by Apple's on-device facial recognition and tagged as "unknown individual, Windsor Park, 4:57 a.m." The dog walker at mile three wore AirPods with cameras that photographed Renée's body in infrared. The house on Loyola had a Ring doorbell that captured her passing at 5:02 and stored the footage on Amazon's servers where it was available to the Austin Police Department through Amazon's Law Enforcement Portal without a warrant. The house on Berkshire had a Nest doorbell that did the same through Google's servers. The house on the corner had a Wyze camera mounted above the garage that recorded every person who walked past and uploaded the footage to a cloud server in China that Wyze's privacy policy disclosed in paragraph forty-seven of a document nobody had read.

Renée could strip herself naked of technology and the neighborhood's technology would still photograph her, track her, catalog her gait, and file her body's signature alongside every other body that moved through the residential WiFi field between 4:50 and 5:40 a.m. The system wasn't on her. The system was around her. The system was the air. And the air belonged to everyone, which meant the surveillance belonged to everyone, which meant opting out was architecturally impossible because your neighbor's doorbell didn't need your consent to photograph your morning. It was the most deliberate decision she made every day.

She'd stopped carrying her phone on runs three years ago. She'd told people she wanted to be "present." The real reason was operational. Renée had spent twenty years in intelligence work, and she knew what phones were before most Americans started asking the question. A phone was a microphone you paid for. A GPS tracker that never died—the betavoltaic cells in modern phones meant they never powered down, never ran out of battery, never offered the mercy of a dead device in a dead zone. Even off, even in airplane mode, a 2030 smartphone's low-power locator beacon continued pinging the satellite network that Apple and Google maintained for their "Find My" ecosystems, broadcasting the device's position to within three meters anywhere on Earth. Renée left her phone on the nightstand next to the Bible, face down, and for forty-seven minutes every morning she moved through the world generating as little data as a human being in 2030 could generate, which was not zero.

Because the streetlights saw her. The NexGrid adaptive LEDs—installed two years ago as part of Austin's smart infrastructure rollout—brightened as she passed and dimmed behind her, their motion sensors tracking her body through the neighborhood with the blind, faithful attention of a system that didn't know her name but recognized her movement signature with 94 percent confidence. And the WiFi saw her. The mesh network that blanketed Windsor Park—emanating from every router in every house she passed, the signals penetrating walls and fences and the dark air itself—sensed her body disrupting its waves. NexGrid's Ambient Identity Resolution system, deployed across Austin as "pedestrian flow optimization," could identify a specific human being by the way their body interrupted a WiFi signal: gait, height, shoulder width, the unique geometry of a person moving through radio space. Renée knew about this technology because she'd read the classified briefings. She knew the system could identify her without a phone, without a face, without a device of any kind. She ran anyway. The running was not about being invisible. It was about being, for forty-seven minutes, less visible than the rest of the day required her to be.

And then the morning the system reminded her that less visible was not invisible. A Tuesday in the second week of October. She'd run her seven miles — varied route, no devices, the Asics and the Casio and the dark streets. She'd come home. She'd showered. She'd picked up her phone from the nightstand to check the time, and the phone showed a notification from the Nextdoor app she'd installed two years ago to monitor neighborhood activity near Amara's apartment: *Your neighbor posted a new safety alert.* She opened it. The post was from a house on Loyola — three blocks from her own. A Ring doorbell video. Timestamped 5:07 a.m. The caption: "Unidentified woman running past my house before dawn. No glasses, no phone visible. Anyone know who this is? Just want to make sure the neighborhood is safe." The video showed Renée. From behind — her Asics, her silhouette, the rhythm of her stride in the streetlight. The camera had captured her from the homeowner's front porch. She was identifiable by body shape. The post had four comments: "Probably just a jogger." "Still, be careful." "I've seen her too, she runs every morning." "Why no glasses though? That's weird."

Renée stared at the phone. Her analog run — her forty-seven minutes of freedom, her daily act of resistance, the one hour when she was less surveilled than any other hour — had been captured by a neighbor's doorbell, posted to a social media platform, and flagged as suspicious because the absence of technology was, in 2030, a reason for concern. She wasn't hiding from the system. She was hiding from her neighbors' systems. And her neighbors' systems didn't need her consent to photograph her, post her, and discuss her in a forum she couldn't control.

She deleted the Nextdoor app. The deletion removed her access. It did not remove the post. It did not remove the video. It did not remove her body from the Ring doorbell's database or the Nextdoor server or the phones of the four people who'd commented. The system didn't live on her phone. The system lived on everyone else's.

And the Waymos saw her. Two passed during her run —ghostly white Jaguar crossovers gliding through the predawn streets on their way to early-morning pickups, no drivers, no passengers yet, just twelve cameras each and three LiDAR arrays per vehicle mapping everything within 300 meters. The cars photographed Renée from the front, the side, and the rear as they passed. They captured her height, her build, her clothing, her approximate speed. The data went to Google. Google kept it. The Waymos didn't care that she was running without a phone. They saw her body and that was enough.

Six miles. The route varied—never the same two days in a row, never a pattern that could be predicted, because Renée understood pattern detection the way a locksmith understands locks. She knew how profiles were built because she'd built them. She knew what data looked like when it assembled into a person—a cloud of points coalescing into a name, an address, a threat score, a human being reduced to a probability. She ran different routes the way a locksmith uses a different key every day. Not because she was hiding. Because she knew what happened to people who were found.

Home by 5:40. The house was clean in the way that military discipline produces cleanliness—not the absence of mess but the active suppression of disorder. Surfaces wiped. Floors swept. The bed made with hospital corners she'd learned at nineteen and never unlearned. There was nothing decorative that hadn't been considered. A small wooden cross above the bedroom door—her mother's, brought from Lagos in 1986. A framed photograph of her parents on the bookshelf, both in Sunday best, both now dead. The Bible on the nightstand, King James, spine cracked at Psalms and Romans and the book of Ruth.

The house had a Nest thermostat because the landlord had installed it before she moved in. It had a Ring doorbell because the property management company required it. It had a WiFi router because a house in 2030 without WiFi was a house without electricity—functionally uninhabitable, socially suspicious, the kind of absence that generated questions she didn't want to answer. Renée had not purchased a single smart device voluntarily. No Echo. No smart refrigerator. No Roomba. No smart television. She owned a 2019 Samsung TV—old enough that its microphone had been disabled by a firmware update she'd manually blocked, its WiFi disconnected, its only function to display whatever she chose to play through a wired HDMI connection. Her colleagues would have found this eccentric. Her colleagues bought every new device the moment it shipped and connected it to every platform available and lived inside the ecosystem the way tropical fish live inside aquariums—comfortably, beautifully, without ever seeing the glass.

She showered. Dressed. Dark slacks, navy blouse with a structured collar, flat shoes she could move in. No jewelry except the Casio. She put on her glasses—Apple AI Glasses, third generation, the plain black Warby Parker frames she wore because the facility required them for the building's biometric access system and because walking through Austin in 2030 without smart glasses marked you as either elderly, impoverished, or deliberately noncompliant, and Renée could not afford to be any of these things. She wore the glasses from the moment she left her front door to the moment she returned. She understood what they were: a camera she'd been socially required to wear on her face. She wore them anyway. The alternative was visibility of a different kind.

The drive to the facility took twenty-two minutes. She drove a 2027 Toyota Camry—chosen because it was the most common car on the road and because its connected-vehicle features, while impossible to fully disable without triggering a dashboard warning, were less invasive than those in newer models. The Camry's Toyota Connected Services still transmitted her location, speed, and route to Toyota's data platform every thirty seconds. The onboard diagnostics system still logged engine performance, fuel consumption, and mileage. But the 2027 model didn't have a cabin camera—those became standard in 2029—and its telematics could be partially obscured by disconnecting the vehicle's eSIM, which Renée had done with a SIM extraction tool and a YouTube tutorial, creating a gap in the data stream that

Toyota's system registered as "intermittent connectivity, service appointment recommended."

She took a different route than yesterday. Left on Berkman instead of right on Airport. Through the Mueller development instead of around it. Small deviations that would register as noise in a pattern analysis, not signal. She drove past four Waymos in seven minutes—each one photographing her car, her plate, her face through the windshield. She drove past a Cruise shuttle at a stoplight. She drove past a NexGrid smart streetlight at every intersection, each one capturing her plate and adding it to the Flock Safety database shared with 8,000 law enforcement agencies. The Camry's disconnected eSIM left a gap in the Toyota telemetry, but the gap was its own signal—a car that should be transmitting and wasn't, an absence the system noted the way a body-sensing network noted the absence of a body in a space where a body usually was.

Her phone—left on the nightstand, not in the car—was still broadcasting to the cell tower on Berkman Drive, its location ping transmitting every twelve seconds to AT&T's network, the ping stored in AT&T's records for a minimum of seven years under the carrier's data retention policy, the records available to law enforcement through a tower dump request that required no warrant, only a court order that judges approved routinely because the legal standard for cell tower data was "reasonable grounds" and "reasonable grounds" was whatever the requesting agency said it was. Renée's phone said she was at home. Renée was on Research Boulevard heading to a facility she couldn't name. The gap between the phone's location and her body's location was the margin she maintained between her digital self and her physical self—the same margin she maintained when she drove to Amara's apartment, the same tradecraft, the same daily act of separation that consumed energy the way running consumed energy, the body's resources diverted to the work of being less visible in a city that was built to see.

The route variations she practiced with such discipline were rendered partially meaningless by the density of the surveillance infrastructure she drove through—but partially was not entirely, and Renée operated in the margins because the margins were all she had.

The facility didn't have a sign. It occupied a low-slung building in a business park near two highways, surrounded by insurance offices, a dental practice, and a FedEx center. The parking lot required a badge. The lobby required a badge and a palm scan. The elevator required a badge and a code. The floor required a badge and a retinal scan. By the time she sat down at her workstation, she had been authenticated six times, each event logged in a system that created a daily timeline of her movements more detailed than any diary she'd ever kept.

"Morning, Renée." Linda, two desks over. Intelligence support. Good at her job, bad at reading rooms.

"Morning."

"New SIGINT brief. Three names on the southwest corridor. All financial."

Renée nodded. She pulled up the classified environment—air-gapped from the internet, running software with no consumer equivalent. The interface was plain. No design aesthetic. Function only. She opened the daily feed—alerts, reports, flagged activities from the interagency pipeline that connected federal, state, and local systems into a single searchable architecture. She scrolled the way a doctor scrolls lab results—quickly, looking for the anomaly.

The scope of what she could see from this chair was something she'd stopped thinking about years ago. With the right query, she could pull the location history of any phone number in the United States going back three years. She could access the Waymo fleet's captured imagery for any street address in any city where the vehicles operated—Austin, Phoenix, San Francisco, Los

Angeles, New York—and view time-stamped photographs of the building, the sidewalk, the people passing, taken from the cameras of autonomous vehicles that had been driving past all day. She could cross-reference facial recognition databases with DMV photographs from forty-eight states. She could see the smart-shoe GPS data that fitness platforms sold to data brokers who sold it to intelligence agencies who processed it into movement profiles. She could pull the Ambient Identity Resolution logs for any WiFi-covered area in Austin—the system that identified people by the way their bodies interrupted radio waves—and trace a person's movements through the city without their phone, without their face, without any device at all. Just a body in space.

The system was vast and it was legal and it was running in this room, accessible to the thirty-two analysts who held the clearances. Renée was one of the thirty-two. She'd earned the trust over twenty years—Army intelligence, two deployments, a transition to civilian analysis her superiors had encouraged because she was precise and discreet and understood the moral compartmentalization the work required. She was good at compartments. The analyst. The churchgoer. The aunt. The woman who read romance novels on a Kindle at midnight—the Kindle she'd kept because it was the one Amazon device she permitted herself, and only because the alternative was paper books, and paper books required bookstores, and bookstores had cameras. Each compartment sealed. Each one holding a version of Renée the others never saw.

The day passed. Briefings. Analyses. The low hum of classified work. At lunch, she ate a commissary salad at her desk—efficient and joyless, the Styrofoam cup of weak coffee confirming what the facility's aesthetic said plainly: the budget went to servers, not people. She checked her personal phone once, in the break room. A text from her mother's nursing home—medication update, routine. She replied with a thumbs-up. She pressed her thumb into her opposite palm—the tell she didn't know she had, the thing her body did when her compartments were under pressure—and went back to work.

At 4:45, she logged out. Badge through every checkpoint in reverse. The parking lot. The Camry. And then—instead of turning left toward Windsor Park—she turned right.

The route to Amara's apartment was never the same twice. Today she drove south on I-35—passing six Waymos, a Cruise shuttle, and a Tesla operating in full self-drive mode with the driver watching a show on the dashboard touchscreen—exited at Riverside, looped through a neighborhood of student housing and taco trucks, and parked two blocks from the complex. She left her phone in the glove compartment, still connected to the cell tower, still broadcasting her location as this parking spot. She walked the rest. Four minutes. No phone, no vehicular data connecting her plate to the address. If anyone queried Renée's location at 5:15 p.m. on this Wednesday, they'd find her parked on a residential street in South Austin. They would not find her at apartment 2C in the Vista Ridge complex, where a twenty-one-year-old woman was opening the door with a smile that undid every compartment Renée had built.

"Auntie."

"Hey, baby."

Amara was five-five and slight in a way that made people underestimate the space she occupied. Dark skin like Renée's. Natural hair in twists she maintained herself on Sundays—a ritual Renée had taught her because Amara's mother, Renée's sister, had returned to Nigeria when Amara was eleven and wasn't here to do it. She wore a cropped t-shirt and jeans and old Converse and a thrifted denim jacket she'd had since high school, and she looked like exactly what she was—a twenty-one-year-old woman from Austin, Texas, who spoke with a Texan drawl and knew every taco truck on East Riverside and had never been to the country on the documents she didn't have.

The apartment was small—a one-bedroom with a kitchen that doubled as a dining room and a living room that doubled as everything else. The walls were thin enough to hear the neighbor's television, a Spanish-language game show whose applause track bled through the drywall at regular intervals. The kitchen smelled of stewing tomatoes and habanero and palm oil—the layered, building heat of jollof rice, the dish Amara made better every week, the flavors deepening as her confidence grew. She was always cooking when Renée came, always something Nigerian, learned from YouTube channels run by women in Lagos who narrated in Pidgin and laughed while they cooked, women who reminded Renée of her own mother and who reminded Amara of nothing, because Amara had no memory of Lagos, only the food. The food was the memory. The food was the country she'd never return to, served on mismatched plates in a one-bedroom apartment in Austin, Texas, and it was the most Nigerian thing about her. Amara's phone—an old iPhone 20, three years outdated, registered to Renée's name—was propped against the salt shaker, the screen showing a woman's hands stirring a pot. Amara cooked Nigerian food for a country she didn't remember, in a city that didn't know she existed, and served it to the only family member who'd stayed.

Amara wore glasses—the budget Apple AI Glasses model, the ones without the AR display, just the camera and microphones and basic Siri functionality. She wore them because everyone wore them and because not wearing them would mark her as someone who couldn't afford them, which was a kind of visibility she couldn't risk. The glasses recorded everything she saw. The phone tracked everywhere she went. The data flowed to Apple's servers, where it sat alongside the data of 1.4 billion other users, undistinguished, unexamined, safe only because the system hadn't been asked to look at it yet.

Renée sat at the kitchen table. She watched Amara stir and felt the tenderness and the terror in equal measure, the two emotions so intertwined she could no longer separate them.

"You good?"

"I'm good, Auntie. Work was fine. Miriam was annoying but she's always annoying."

"Same route this week?"

"Yes." A flicker of exasperation. "Same route. Different bus times like you said. I varied the mornings."

"Good."

"Can I ask why you're asking more?"

Renée looked at her niece—this woman who had grown up in this city, graduated from its schools, walked its streets, and was, in every observable way, from here. The only thing separating Amara from the students she'd sat next to in homeroom and the coworkers she ate lunch with was a database entry. A record. An absence of the right paperwork in a system that defined belonging by documentation and enforced it with the same data infrastructure Renée spent her days operating. Palantir's ImmigrationOS was running on a server in the same building where Renée checked her SIGINT briefs. The system that protected the country and the system that would deport her niece drew electricity from the same outlet.

"I just want you to be careful."

"I know."

"Call me when you get home tomorrow."

"Auntie. I always call."

Amara set a plate in front of her. The jollof was good —better than last time, the rice smoky, the peppers right. Renée ate and said, "This tastes like home," and Amara smiled and didn't say what she was thinking, which was that she wouldn't know, because she'd never been.

They ate together. They talked about small things—Amara's coworker, a show, a song Amara played from her phone. The conversation generated no intelligence. No flags. No keywords. It was two women eating rice in a kitchen, and the most important information exchanged was not in the words but in Renée's presence, which said: I am here. The system that employs me is the system that hunts you, and I have chosen you, and I will keep choosing until the choice is taken.

She pressed her thumb into her palm under the table. She asked for seconds.

The drive home. North on Pleasant Valley, west on 7th, through campus. A third route. Three Waymos passed, photographing the Camry each time. In her driveway, she sat in the car for two minutes, engine off, the onboard system transmitting its final location ping to Toyota's servers. Then she went inside.

The house was dark and clean. She changed into a t-shirt and cotton pants. She made tea. She took off the glasses and set them on the dresser, cameras and microphones finally dark, the betavoltaic cell still running behind the glass, the device still alive even when she wasn't wearing it, a small machine patience on her dresser waiting for morning.

She sat on the edge of her bed and opened the Bible. Romans 13. *Let every soul be subject unto the higher powers.* She'd been reading this passage for a week, turning it the way she turned intelligence, looking for the angle that made it mean what she needed. Be subject to authority. She was subject. She wore the badge. She took the oath. And every Wednesday she drove to South Austin and committed federal offenses to protect a girl the authority would destroy.

She closed the Bible. She picked up the Kindle—her one permitted Amazon device, the one screen she allowed herself in bed. Three books in rotation: a romance novel about a woman being held by a man who deserved her, *RESIDUAL* by Geoffrey Arbuckle—the second Nora Cafferty thriller, the one that had fictionalized the Epstein files before the congressional hearings caught up, the one where three names in the novel matched three names on the subpoena list and the author had said "I made it up" and the public had said "you knew"—and the King James Bible she'd just closed. She read two pages of the romance. She read a chapter of *RESIDUAL*. Nora Cafferty was in a server room in Zurich, pulling financial records that connected a dead man to a living senator. Renée set the Kindle down. The fiction was too close to the work. The work was too close to the fiction. She picked up her phone and opened a music app, and for ten minutes, while the tea cooled, she played Yoruba worship songs at a volume so low they were almost inaudible—her mother's music, hymns from a church in Lagos she'd attended until she was twelve, melodies that bypassed every compartment and found the version of Renée that existed before the uniform, before the clearance, before the divorce. The music lived in a corner of her streaming account that the algorithm didn't know what to do with—it didn't match her behavioral profile, didn't correlate with any demographic. It sat there like an artifact from a person the system hadn't been built to understand.

Outside, a Waymo passed the house on Berkman Drive. Its cameras captured the dark windows, the unlit porch, the car in the driveway. It didn't know who lived here. It added the photograph to the fleet's continuous map of Austin, the city rebuilt in data every hour of every day, and kept driving.

Renée pressed her thumb into her palm. She listened to her mother's music. She slept.

-e

***

CHAPTER 4

The bar was loud enough that Elena Ríos had to lean in to hear Marco describe the end of American public education, which he was doing with the cheerful despair of a man who'd been watching it happen for nine years and had recently stopped believing it could be reversed.

"So this kid—twelfth grader, smart kid, I've had him since tenth—he's wearing the Apple AI Glasses." Marco paused to drink his Modelo. "Not the cheap ones. The Warby Parker ones with the full AR display. He's sitting in my class during the exam, and he's just—" Marco looked up slightly, mimicking the gesture. "Looking up. Like this. Reading the answers off the inside of his lens. Not even hiding it."

"What'd you do?"

"What could I do? I told him to take them off. He said they were prescription. They weren't prescription—his mom bought them at the Apple Store in the Domain three weeks ago, I know because she posted the unboxing on her Instagram. But now I'm in a conversation about medical accommodations and district policy and whether I can legally require a student to remove corrective eyewear, and meanwhile the kid in the next row is using his AirPods to whisper questions to Siri Sense and getting answers fed directly into his ear through bone conduction, and the kid behind him submitted an essay last week that was either the best thing a seventeen-year-old has ever written about the Reconstruction or was generated by Claude in four minutes. And I can't tell which. I literally cannot tell."

Elena laughed. She couldn't help it. Marco told these stories with the timing of a man who'd been workshopping them during the drive home, and the result was that his descriptions of institutional collapse sounded like stand-up. Thirty-three, five-eleven, solid through the chest from the soccer he used to play, dark hair that needed a cut, warm brown skin, the chalk dust still visible on his cuff. He wore his glasses pushed up on his forehead—the Meta Ray-Ban AR 4, the teacher-budget model—in the gesture that had become the 2030 equivalent of hanging up the phone: pushing your glasses up meant you were off the record, not recording, present. Some people did this. Many didn't bother. Marco always did. It was one of the things Elena loved about him—his instinct that some moments should belong to the people in them and not to the cloud.

They were at Pueblo Viejo on East Sixth, a taco bar with a corrugated-metal ceiling, Edison bulbs strung between exposed ductwork, and the smell of charred corn and lime and pork fat that hit you at the door like a welcome mat made of appetite. A cumbia track competed with the crowd noise—thirty or forty people at wooden tables, the Thursday-night regulars, people who came here because the Al Pastor was $4 and the mezcal was honest and the vibe was the old Austin that was disappearing block by block under glass towers and Waymo stations.

The cumbia was playing through an Amazon Echo Studio mounted on a shelf behind the bar—the speaker the owner, Miguel, had bought because the sound quality was good and the Alexa integration let him control the playlist by voice while his hands were in tortilla dough. Miguel loved the speaker. The speaker loved Miguel back—or rather, the speaker listened to Miguel and every other voice in the restaurant with the always-on microphone that Alexa's privacy settings described as "wake word detection processing" and that functioned, in practice, as a continuous audio monitor sampling the ambient sound in three-second buffers, processing the audio through Amazon's speech recognition engine, and discarding the samples that didn't contain the wake word. The samples were discarded. The metadata—audio volume levels, number of distinct voices detected, ambient noise classification, duration of activity—was retained and uploaded to Amazon's analytics platform, where it was processed alongside the audio metadata from 500 million other Echo devices worldwide to build a behavioral model of human activity patterns that Amazon sold to advertisers and shared with its "service improvement partners." The Echo in Pueblo Viejo knew how many people were talking at any given moment. It knew the noise level at 9 p.m. versus midnight. It knew when the crowd thinned and when it surged. It didn't know Elena and Marco were discussing the collapse of public education. It knew two voices at table seven had been in sustained conversation for forty-seven minutes, which was above the venue's average interaction duration, which made them "high-engagement patrons," which was a data point.

The Samsung smart TV behind the bar showed the Longhorns game on mute—the screen that Miguel had mounted last year because customers wanted to watch sports and the Samsung Frame TV had been on sale at Costco and the picture quality was beautiful. The TV's built-in microphone and camera were in "ambient mode"—not actively recording, but passively sensing the room's occupancy through its "Presence Detection" feature, the same feature that powered the TV in Daniel's apartment and the one in the Holloway living room. The TV knew how many bodies were in the restaurant. The TV knew which direction they faced. The TV uploaded this data to Samsung's SmartThings platform through the restaurant's WiFi, and the data joined the behavioral profiles of every other Samsung TV in every other bar and restaurant and living room and bedroom, and the aggregate was a map of human attention that Samsung sold to content providers and advertisers and that nobody in Pueblo Viejo—not Miguel, not Elena, not Marco, not the thirty regulars eating $4 tacos—had consented to because the consent was in the terms of service Miguel had agreed to when he'd plugged in the TV and the terms were forty-three pages and the AGREE button was green and the music was good and the tacos were ready.

A touchscreen ordering system was embedded in the table surface, scanning their faces when they sat down—not for identity, supposedly, just for "personalized menu recommendations." Elena had ordered here eleven times. The table knew she preferred mezcal neat and the Al Pastor plate. The table also knew Marco's Modelo preference, their average visit duration (94 minutes), and their typical check split (50/50). Elena knew all of this. She sat at the table anyway. The mezcal was smoky and warm and tasted like Oaxaca and the company was better and the alternative was never going anywhere, which was its own kind of surveillance—the absence of data was also data.

"But that's not the thing," Marco said, leaning forward. "The thing is what happened with the girl."

"There's always a girl."

"Sophomore. Fifteen. She took a photo—I don't know what kind, I don't want to know—and sent it on Snapchat. The friend screenshots it. By fourth period it's on every device in the building. Not just phones—the glasses too. Kids are AirDropping it through their Apple AI Glasses in the hallway. The image is in thirty kids' cloud backups before lunch. The girl is in the bathroom crying and her friends are standing outside the door and nobody comes to a teacher. They handle it on their devices. They're having a crisis inside the ecosystem about something that happened inside the ecosystem and the resolution is going to happen inside the ecosystem, and I'm standing in my classroom talking about the Treaty of Versailles to thirty kids who are physically present and mentally on a different planet."

He said this without anger. That was the thing about Marco—his bewilderment was genuine. He wasn't performing outrage. He was describing what he saw with the uncomprehending honesty of a man who'd become a teacher because he believed the classroom was sacred space, and the devices had broken that, and he didn't have the vocabulary for what had replaced it.

Elena did. She heard "image AirDropped through AI glasses in a hallway" and her brain translated instantly: image stored on Snap's servers and in Apple's iCloud for every device that received it, metadata logged, the girl's face and body replicated across infrastructure she had no control over, each copy generating data that Snap's analytics processed for engagement metrics, a fifteen-year-old's worst moment indistinguishable from content. The glasses had recorded the hallway where the sharing happened. The school's WiFi had logged every device involved. The data existed to identify exactly who shared the image and when—but no one at the school knew how to query it, and the platforms had no obligation to tell them.

She didn't say any of this. She'd learned not to. Early in their relationship she'd tried translating Marco's stories into surveillance architecture and his eyes had glazed in a way that told her she'd crossed the line between interesting and exhausting.

"That's brutal," she said.

"It's Tuesday. That's just Tuesday."

She took a drink of the mezcal—Wahaka, neat, the good stuff she allowed herself once a week because the budget was what it was. Five-four, dark hair pulled back in a clip, wearing a thrifted blazer over a Pavement t-shirt, jeans, boots that had walked through enough government buildings and public records offices to qualify for retirement. Her Apple AI Glasses—the cheapest model, no AR display, just the camera and microphones she couldn't avoid in 2030—sat on her face like a compromise she'd made with the century. She wore them because not wearing them in professional settings was like not having a phone in 2010: possible but career-limiting. Her messenger bag sat beside her, heavy with a laptop and two notebooks and the accumulated weight of a career spent asking questions institutions didn't want answered.

"Can I ask you something? The glasses kid. Did the school address the technology, or just the cheating?"

"What do you mean?"

"I mean, did anyone say, this kid is wearing a camera with a six-microphone array in a classroom full of minors? Every face in your room was captured and processed through Apple's neural engine. Did anyone flag that?"

Marco looked at her. "No. They wrote him up for academic dishonesty."

"Right."

"You're doing the thing."

"What thing?"

"The thing where I tell you a school story and you turn it into a surveillance story."

She smiled. He wasn't wrong. "Sorry."

"Don't be. Just—sometimes I want to tell you about the kid with the glasses."

"Tell me again. I'll just listen."

He smiled. The smile was the reason she'd said yes to the second date and the eighth month of whatever this was—a smile without agenda, without calibration, the smile of a man genuinely glad to be here. In a city where everyone Elena interviewed was measuring every word, Marco said what he meant.

They ordered another round through the touchscreen table. He told her about the AI-generated essay—a student who'd used a prompt chain sophisticated enough that the AI detection software couldn't flag it, producing a paper on Reconstruction that was structurally brilliant and emotionally hollow. "It had all the right words in all the right places," Marco said, "and it meant nothing. I could feel that it meant nothing. But I couldn't prove it meant nothing." Elena thought about the collapse of authenticity across every domain simultaneously—education, journalism, photography, identity—and said "that's really interesting, babe" and meant it and also meant the seven things she was thinking behind it.

Then Marco got quiet—the quiet that meant Jaylen. Jaylen Torres, class of 2024, AP History, the best paper on the Treaty of Versailles Marco had ever received from a student, handwritten, genuine, the last paper Marco graded that he was certain no AI had touched. Jaylen had enlisted after graduation. Deployed to the Persian Gulf in March 2026. Killed by a drone strike in the Strait of Hormuz theater on April 19th—two weeks after the war started, six weeks before the ceasefire. Marco kept the paper in his desk drawer at school. He didn't talk about Jaylen often. When he did, the talking sounded like this: silence, then a swallow, then "the same cameras, Elena. The same systems. They used them over there and now they're using them here and nobody sees it because they're not looking." He didn't have the vocabulary. He had something better—the grief of a teacher who'd watched a student walk out of his classroom and into the infrastructure.

They split the check. Marco's card. Elena's card. The table processed both, logging the split, the tip percentage (22% each), the timestamp. He kissed her outside—quick, warm, his hand on the side of her face, the moment captured by the bar's exterior camera and by the Flock Safety license plate reader on the utility pole across the street and by a Waymo that glided past on East Sixth, its twelve cameras photographing every face on the sidewalk.

Elena ordered a Waymo home. The autonomous Jaguar arrived in three minutes—no driver, no small talk, the interior clean and cool, the seats adjusting to her body as she sat. She gave the destination through the car's voice interface—her South Austin address, spoken aloud, recorded by the vehicle's interior microphone, the destination logged in Google's ride database alongside every other Waymo trip she'd ever taken. The interior camera—mounted above the rearview mirror, wide-angle, always active—captured her face, her posture, her expression, the phone in her hand, the angle at which she held it. The seat sensors measured her weight and weight distribution. The climate system adjusted to her body temperature. The Waymo's interior was a sensor suite disguised as a luxury sedan—every surface collecting, every system reporting, the ride classified in Google's database as "User ELR-4792, pickup: East 6th and Sabine, destination: South First and Oltorf, duration: 14 minutes, passenger status: single occupant, mobile device active, emotional classification: neutral." The car drove itself across the river, past the convention center, through the gentrification gradient that ran from the new condos on Rainey Street to the older apartments near Oltorf.

Her phone connected to the Waymo's WiFi automatically—a handshake the device performed without asking, the phone and the car exchanging data in the background the way two machines shake hands: device ID, carrier, operating system, installed apps. The Waymo's WiFi now knew every app on Elena's phone. The phone knew every sensor in the Waymo. The handshake lasted 0.4 seconds and generated a data packet that would be stored on Google's servers for seven years—the default retention period, buried in the terms of service Elena had agreed to when she'd created her Waymo account.

Her credit card—the Visa she used for everything because cash was increasingly difficult and decreasingly accepted—had been charged $11.47 for the ride. The charge appeared simultaneously on her bank's transaction log, on Visa's merchant processing platform, on Google's payments database, and on the financial analytics platform her bank shared data with under a partnership agreement disclosed in a quarterly privacy update Elena had received via email and not opened. The $11.47 described a woman leaving a taco bar at 11 p.m. on a Thursday. The $4.00 Al Pastor plate and the $14.00 mezcal from Pueblo Viejo were logged separately—cash register to Square payment processor to the restaurant's POS analytics to the data broker that aggregated restaurant spending patterns and sold them to commercial real estate analytics firms and health insurance actuarial platforms. The mezcal was categorized as "alcohol purchase, weekly frequency: 1, consistent with moderate consumption." Elena's health insurance premium had not been adjusted. Not yet. The data was patient.

Elena sat in the back of a driverless car on a Thursday night and watched the city pass through tinted glass and thought about the fact that she was sitting inside a mobile surveillance platform whose cameras were, at this moment, mapping every face on every sidewalk between East Sixth and South First, feeding the data to Google and, through NexGrid's traffic integration, to the same municipal surveillance system she spent her days investigating.

She knew this. She got in the car anyway. Because what else was she supposed to do—walk home to South Austin at 11 p.m. to avoid generating a data point? The system won the moment it became easier to comply than to resist. That was the whole game. The system didn't force compliance. It made compliance frictionless and resistance exhausting. Her bank—Chase, the one she'd had since college—had closed the South Austin branch two years ago, consolidating into "digital-first banking" that required every transaction to flow through the app, every deposit to be photographed by the phone's camera, every payment to generate a data trail. She'd tried to keep a cash budget—$200 a week in twenties, enough for tacos and coffee and the transactions she wanted to keep off the ledger. But cash was dying. Three restaurants on South First had gone cashless in the past year. The laundromat required an app. The parking meters accepted only digital payment. Even Pueblo Viejo—the taco bar that still felt like old Austin—had replaced its cash register with a touchscreen that defaulted to card and required three extra taps to process cash, the friction designed to nudge you toward the tracked path the way a cattle chute nudges the animal toward the processing floor. The system didn't ban cash. It made cash inconvenient. It made analog difficult. It closed the branches and the windows and the physical spaces where humans transacted with other humans without a server between them, and it replaced them with apps that were faster and easier and that logged everything, and the logging was the price and the price was invisible because the service was free and the free was the most expensive thing you'd ever agree to.

Her apartment was the opposite of Renée's house and Daniel's unit. Where Renée maintained control and Daniel maintained absence, Elena maintained chaos. The living room was a newsroom compressed into four hundred square feet. The whiteboard on the wall—red and blue string connecting printouts and Post-it notes, what Marco called "the conspiracy board." Three laptops on the desk, each running a different VPN, each logged into different accounts. Stacks of FOIA printouts in plastic bins labeled by date and agency. Books on surveillance law, on data privacy, on intelligence overreach—spines cracked, pages flagged with colored tabs. A coffee mug with dried espresso. Two empty Topo Chico cans. The remains of a life that consumed information the way other lives consumed food—constantly, compulsively, never quite sated.

She dropped her bag and sat and opened the laptop and the world she'd held at arm's length during dinner rushed back.

She was working on a FOIA request to the City of Austin regarding municipal contracts for smart infrastructure. Filed eleven weeks ago. The city's PIO had sent two extensions citing "volume and complexity"—bureaucratic for "we don't want to give you this." She'd filed an appeal. Meanwhile, she'd built a timeline from public procurement records—council agendas, budgets, vendor registrations—showing infrastructure spending that accelerated sharply in 2027 and pointed to a single prime contractor whose name appeared on sixteen line items across four departments.

NexGrid.

She didn't know what NexGrid was yet. Not fully. She knew it was a smart city platform. She knew it held the Austin contract. She knew the contract had expanded three times. She knew the CEO was Catherine Hale, background in defense contracting, keynotes titled things like "The Visible City: Infrastructure as Insight." Elena had watched two keynotes and taken four pages of notes and felt the specific chill of hearing someone describe mass surveillance in the language of civic improvement while the audience applauded. She knew that NexGrid's platform now integrated the Waymo fleet's camera data—the autonomous vehicles feeding real-time street imagery to the same system that managed the city's fixed surveillance infrastructure. She'd ridden home tonight inside one of the system's eyes.

But she didn't have the inside. She had the public face. The press releases. Not the architecture—the data flows, the endpoints, what NexGrid was actually doing with the information from every sensor and camera and WiFi node and autonomous vehicle in the city. She needed a source inside the company, and she didn't have one.

She checked The Circuit's dashboard. Forty-two hundred subscribers. Loyal, engaged, niche. Not enough to pay rent. She freelanced to cover the gap—Texas Monthly, wire stories, research contracts. Rent due in nine days. A freelance check fifteen days late. Her savings balance was a number she tried not to look at because looking at it made the FOIA appeals feel less like journalism and more like a hobby she couldn't afford. On the bookshelf beside her desk, between the FOIA printouts and a stack of municipal procurement records: *The Forgotten Standards* by Geoffrey Arbuckle, a nonfiction book she'd bought during a 2 a.m. research rabbit hole six months ago, and *Complicit*, the Nora Cafferty thriller that every journalist in America had read after the Meridian hearings when fiction turned out to be reportage. Beside them, a framed front page from The Circuit's coverage of the Iran-theater surveillance pipeline—her three-part series from 2027, "The Cameras

Come Home," about how the Palantir targeting systems and AI-driven facial recognition deployed against Iran had been reshored into American police departments within eighteen months. Nobody had read the series. Forty-two hundred subscribers. But the work was on the shelf and the shelf held the architecture of everything she'd been building toward without knowing it.

She worked until 1:40 a.m. CNN muted on the center laptop—a Rubio administration press briefing on proxy containment, the chyron scrolling: PRESIDENT ADDRESSES NATION ON ASYMMETRIC THREATS — the same word, "threats," that NexGrid used in its confidence scores. Three new Post-its on the whiteboard. A follow-up on the FOIA appeal. Two new entries on the NexGrid timeline. Then the wall—the hour when her eyes burned and the information stopped assembling into meaning.

She closed the laptop. She opened YouTube on her phone. The algorithm—which knew her search history, her watch time, her pause patterns, and the fact that she watched cooking videos between 1:00 and 2:00 a.m. with 78% consistency—auto-played a woman in a kitchen somewhere in Oaxaca, hands working masa, narrating softly in Spanish. The camera was close on the hands. The sound was dough and a quiet voice and the hiss of a comal. Elena watched for eleven minutes. Her shoulders dropped from somewhere near her ears.

She didn't watch cooking videos because she cooked. She barely cooked. She watched them because the sounds—the chopping, the narration in a Spanish closer to her grandmother's than her own—did something to her nervous system that no app had managed. They took her back. Not to a memory but to a feeling—the feeling of a kitchen where someone was making something for you, where the labor was love and the love was food and neither required a login.

Her grandmother's photo was taped inside her laptop case. Not in the cloud. Not backed up. A physical photograph, faded, of a woman outside a house in the Rio

Grande Valley with her hand on a chain-link fence and the sun behind her. The only image in Elena's life that existed in a single copy. If she lost it, it was gone. That precarity was what made it precious. Everything else was backed up, synced, permanent. The photo could be destroyed by a glass of water. She kept it because some things should be fragile. Because permanence was the surveillance state's promise, and it was a lie dressed as a gift.

The Valley came back the way it always did at this hour. Not as narrative but as sensation. The heat. The checkpoints where Border Patrol waved some cars through and stopped others, and even as a child she'd understood the selection wasn't random. The white aerostat balloons floating above the landscape like permanent clouds, watching the desert with cameras that could read a plate from fifteen thousand feet. She'd grown up inside a surveillance state before she had a word for it. The border was where America tested its tools before deploying them everywhere else. The Waymo fleet. The WiFi sensing. The facial recognition. The predictive algorithms. Elena had known all of this in her body since she was eight years old, standing in the back seat of her mother's Corolla, watching the balloon through the rear window as they drove north.

She didn't talk about the Valley. Not to Marco. Not in articles. It was the engine underneath everything—the reason she filed FOIAs at 1 a.m. and tracked contracts and read procurement documents nobody else found interesting. The Valley was why. But the Valley was hers.

She reached for her phone. She almost called her mother—asleep since nine, living on a schedule of garden and dog and evening news and rosary, rhythms the surveillance economy hadn't disrupted because they didn't generate enough data to be worth disrupting. Elena wanted to hear her voice. Not for information. For the sound of a world where the worst thing being monitored was whether the tomatoes needed water.

She put the phone down. She closed her eyes.

On the nightstand, the phone sat in the dark, screen black, microphone live. The glasses sat beside it, cameras and microphones in standby, the betavoltaic cell still running, the device still alive. Outside, a Waymo glided past her building on South First, its cameras capturing the apartment complex, the parked cars, the streetlight adjusting for no one. The car didn't know Elena was inside, building the case against the system it served. It photographed her building and kept driving.

She slept. The city didn't.

-e

***

CHAPTER 5

The anomaly appeared at 2:17 p.m. on a Wednesday, in a routing log that Daniel Yun had been reviewing for eleven minutes as part of a quarterly performance audit he'd been assigned because he was thorough and available and because nobody else on the infrastructure team wanted to do it.

He almost missed it. The log was dense—thousands of lines of packet routing data, each entry recording the origin, destination, protocol, timestamp, and data classification of a transfer within NexGrid's Austin platform. The platform processed 4.7 terabytes per hour from 12,000 sensor nodes, 340 Flock Safety license plate readers, 4,200 Waymo vehicle integrations, 89 acoustic arrays, the citywide WiFi mesh, and the Ambient Identity Resolution system that tracked human bodies through radio waves. All of it flowed through routing logs that looked like an endless grid of alphanumeric characters in monospace font on a black background. Reading them was like reading the ocean by staring at the surface—you had to trust your pattern recognition to flag the deviation before your conscious mind understood what it was seeing.

Daniel had spent seven years learning to read these surfaces. He scanned routing logs the way a cardiologist reads an EKG—quickly, trusting rhythm, waiting for the wrong note.

The wrong note was a single line. A data packet originating from Sensor Cluster 7 in East Austin—the dense array of traffic cameras, device proximity sensors, WiFi scanning nodes, Waymo fleet integration feeds, and the body-sensing system that identified pedestrians by gait through radio-wave disruption—routed through the standard aggregation pipeline, anonymized at the first processing layer, and then forwarded to an endpoint designated by a sixteen-character alphanumeric string he didn't recognize.

He stopped scrolling.

In NexGrid's architecture, every endpoint had a name. The city's traffic center was TXDOT-ATX-01. The public safety dashboard was AUSTPD-RTCC. The Waymo integration was GOOG-WMO-ATX. The naming convention was standardized, documented, boring. The endpoint in the log was none of these. Sixteen characters, no hyphens, no agency prefix. It looked like a hash. It looked like something designed not to be readable by human eyes.

He searched the architecture documentation—the master reference he'd helped write. No results. He searched the integration protocol registry—every external system NexGrid was authorized to communicate with. Nothing. He checked the Waymo specs, the law enforcement feeds, the federal data-sharing protocols. The endpoint existed in the live infrastructure but not in any document that described it.

He sat back. Around him, the fourth floor hummed with forty engineers. Ryan-or-Brian was on a video call, Meta Ray-Ban AR 5 pushed up on his forehead, gesturing at a screen share nobody in the room could see but the person on the other end was watching through their own glasses, their own camera, the interaction happening in the air between two pairs of lenses with no screen involved. Priya, two seats over, ate an apple—a Honeycrisp, the specific crunch audible across two workstations—and reviewed code through her Apple AI Glasses, AR annotations overlaying her Dell monitor in a layer of information visible only to her, the code hovering in space like a hologram in a film that had become real so gradually nobody noticed it was remarkable. The La Marzocca espresso machine hissed in the break room. The Bevi sparkling water dispenser tracked consumption by employee badge. The SnackNation wall—a curated selection of protein bars, dried fruit, and artisanal chips restocked weekly by a vendor that logged which items were taken and by whom—showed three empty hooks where the Siete chips had been. The HVAC whispered.

NexGrid's Austin office was, by design, a space that discouraged urgency—natural light from floor-to-ceiling windows that the Schott smart glass could dim on voice command, real plants maintained by a botanical service, a meditation room with a Calm app subscription terminal that nobody used but everyone pointed to during recruiting calls, and a rooftop patio with a view of the Domain's curated roofline that made the campus look like a place where good things happened. The Teramind workforce analytics platform logged Daniel as "workstation active, application: system_logs, session duration: 11 minutes." The ambient occupancy sensor in the ceiling tile above his desk registered his body heat as present. His Apple Watch Ultra 5 recorded 68 bpm, baseline, normal.

Everything here was fine. Everything was working. You are part of something that works. Daniel stared at his screen and felt the first vibration of something he couldn't name—an engineer's instinct for disorder, a wrong note in a song he'd helped compose.

He thought about asking Priya. Three years at NexGrid, part of the original Austin build. She'd know if this was a legacy endpoint—a test environment never decommissioned. Systems accumulated dead endpoints the way houses accumulated junk drawers. It was probably that.

He didn't ask. The endpoint wasn't dead. It was active—processing data in real time, receiving packets at regular intervals, fully integrated into the live architecture. This wasn't forgotten. This was maintained.

He should have flagged it. Opened a ticket. Escalated to the integration team. The process existed for exactly this situation. He'd documented it himself.

He didn't flag it.

Instead, he opened the raw transport data—the deeper layer showing not just destination but routing path. The packet had traveled from Cluster 7 through the standard pipeline: aggregation, anonymization, classification. Then, at the classification layer, it had been duplicated. One copy continued to TXDOT-ATX-01 —the documented destination. The other was stripped of its anonymization header and forwarded to the unknown endpoint.

Daniel read this three times.

The anonymization layer was the wall. The thing NexGrid pointed to in every council presentation, every stakeholder briefing, every press release. "We collect aggregate data. We do not track individuals." The anonymization layer was what made the system legal. It was the difference between urban intelligence and surveillance. Without it, the data from 12,000 sensors and 4,200 Waymo camera arrays was raw, individualized, identifiable—device IDs, WiFi body signatures, facial captures from autonomous vehicles, location coordinates with timestamps, biometric gait patterns from the Ambient Identity Resolution system—flowing from civilian infrastructure to an endpoint that didn't appear in any documentation NexGrid had ever produced.

He took a screenshot.

Reflexive—an engineer's instinct to document. His thumb moved before his brain caught up. The screenshot captured the routing log, the undocumented endpoint, and the evidence that the anonymization layer had been deliberately circumvented by a protocol someone had built into the system on purpose.

He closed the window. Cleared his cache. Opened a performance dashboard. Stared at it.

His Apple Watch recorded the spike: 68 to 94 bpm at 2:23 p.m. The watch flagged it as "elevated resting heart rate, possible stress event" and synced to iCloud, accessible to his insurance provider. His AirPods Pro 6, still in his ears, confirmed the elevation through ear-canal sensors. Two devices on his body knew he was afraid before he'd identified the feeling as fear. The devices didn't understand why. They logged the anomaly the way NexGrid logged its own anomalies—precisely, dispassionately, without comprehension.

He sat at his desk for the rest of the afternoon. Three o'clock standup: "On track, no blockers." His team lead nodded and moved on. Priya, two desks over, was on a call about weekend plans — she was going to the Brooks Rosser concert at Moody Center, the Idol kid from 2026 who'd become the voice of something Daniel couldn't name, an authenticity the algorithm couldn't manufacture. She'd asked Daniel if he wanted to come. He'd said "maybe" and meant "no" and she'd heard both and hadn't pressed. Slack messages answered. A minor code update pushed. His glasses recorded everything he looked at. His watch recorded every heartbeat. His Nike shoes tracked every step between his desk and the break room. Teramind noted his productivity metrics within normal range.

The surface was undisturbed. Underneath, the screenshot sat on his phone like a coal.

He thought about the East Austin heat map. The corridor that ran warmer—more sensors, more cameras, more Waymo passes per hour, more WiFi body-sensing nodes, more data flowing from the neighborhoods where the median income was lowest and the people had the fewest resources to discover what was happening. The corridor had been expanded in 2027, after the Iran War, under emergency authorities that Congress had passed during the Strait of Hormuz crisis and that were supposed to sunset six months after the ceasefire. They hadn't sunsetted. They'd been extended, then folded into NexGrid's permanent municipal contract, the surveillance technology that had tracked insurgents in Tehran now tracking residents in East Austin with the same cameras, the same algorithms, the same confidence scores. He'd written the performance summary. He'd used the word "optimization." The system was functioning within design parameters. The parameters included a direct, unmasked feed—Waymo facial captures, WiFi body signatures, device proximity logs, biometric gait data—routed to an endpoint that didn't exist in any documentation, through a path that bypassed every privacy protection NexGrid had promised the city.

He thought about this. Then he stopped because the alternative was walking into his team lead's office and saying words that would end his career, trigger a visa review, cut the income that funded his parents' treatment, and place him—a foreign national with access to classified municipal infrastructure—at the center of a response he could not survive.

아무것도 아닐 수 있어. It might be nothing.

He knew it wasn't nothing.

At 5:20, badge out. Elevator. The walk through the Domain in the long October light—the sun low enough to turn the glass facades amber, the shadows stretching across Rock Rose Avenue, the air carrying the first hint of evening cool that would settle into the fifties by midnight. A busker with an acoustic guitar was playing something by the fountain, his open case collecting tips and his smart glasses recording every face that paused to listen. Past the Lululemon whose window cameras tracked Daniel's reflection, the Tesla showroom whose scanner read every plate, the Whole Foods whose cameras had logged his face 247 times in fourteen months without him ever buying anything. The streetlights adjusted—brightening for his body, dimming after it. Three Waymos in two blocks. The grackles were roosting in the live oaks along the walkway, hundreds of them, their evening chorus a cacophony of clicks and whistles and electronic-sounding shrieks that rose from the trees like static, the birds settling in for the night while the machines beneath them never settled at all.

The walk felt different now. He'd walked this route 294 times in fourteen months and it had always felt like walking. Today it felt like being processed. The streetlight on Rock Rose wasn't a streetlight. It was a sensor node feeding unmasked data to an endpoint that didn't officially exist. The Waymo on Domain Drive wasn't a car. It was a mobile camera platform connected to the same pipeline. The WiFi from the coffee shop wasn't connectivity. It was a net sensing his body's shape through radio waves, the Ambient Identity Resolution system reading the geometry of his shoulders and the cadence of his walk and matching it against the profile labeled DY-0947.

He was walking through a net he'd helped weave, and the net was finer than anyone outside that building knew.

His apartment. He changed clothes. The Samsung refrigerator's touchscreen suggested a recipe. He ignored it. He opened the refrigerator and took out the jjigae and put it on the stove. The act of heating his mother's soup—stirring it, watching the surface break and reform, smelling the doenjang and the garlic and the gochugaru—was the only thing his hands could do that didn't feel like holding a detonator.

His hands shook. They'd been shaking since Tuesday — a fine tremor, barely visible, the kind a surgeon would notice and a coworker wouldn't. He gripped the spoon tighter. The tremor was worst when he held things in front of people — the coffee mug at standup, the badge at the checkpoint, the chopsticks in the café where he no longer ate because eating in front of colleagues required steady hands and his hands were no longer steady. He ate at his desk now. Alone. The sandwich in one hand and the other hand in his lap where nobody could see it tremble. He'd lost seven pounds in three weeks — the navy chinos that had fit in September now hung at his hips, the gray pullover looser through the chest. His sleep had collapsed into fragments: two hours, then awake staring at the ceiling while the LG television photographed him in infrared, then one more hour of dreams in Korean — his mother's kitchen, the Jagalchi Market, the highway between Busan and Gimhae — then awake again with his hand reaching for the phone that held the screenshot. The Watch logged "severe sleep deficit" every morning. He dismissed the notification. He dismissed everything now. The system was concerned about his health. The system was the thing destroying his health. The irony was a room he lived in.

The Korean was bleeding through. Yesterday in the break room, Priya had asked if he wanted coffee and he'd said "괜찮아요" — *I'm fine* — before catching himself and switching to English. She'd looked at him for a beat too long. He'd smiled. The smile was the mask. The mask was the thing that kept him employed, that kept the visa active, that kept the $400 flowing to Busan. The mask was cracking. He could feel the cracks the way you feel a filling loosening in a tooth — not pain exactly, but the knowledge that the structure was failing, that the thing holding the surface together was eroding from the inside, and that the erosion was accelerating, and that the moment the mask broke completely would be the moment someone at NexGrid looked at Daniel Yun and saw not a competent engineer but a frightened man who knew too much and was shaking.

He hummed while he stirred. A Korean ballad from a highway in 2003—his father's song, about a man who returns home to find that home has changed. The melody drifted in and out of tune. It existed on no streaming platform. It lived only in his father's voice and now, faintly, in Daniel's own. The one sound in his life that had never been recorded, never logged, never processed. It belonged to a car on a highway in South Korea and to a kitchen in Austin and to nothing in between.

He ate standing at the counter. The soup was wrong. Too much water. His mother would have known by the color. She would have adjusted without measuring—a pinch, a pour, a judgment made by a lifetime of practice that no algorithm could replicate. He ate it anyway. Every imperfect spoonful was a line thrown back to Busan, to a kitchen where a woman was sleeping right now and would wake in a few hours and check the blue dot on her Samsung Galaxy Glass 3 and see her son at home, where he should be, doing nothing unusual.

He washed the pot. He sat on the couch. He did not turn on the game. He did not open Hinge—the dating app whose algorithm had been suggesting the same rotation of women for three months, each profile a composite of data points (age, location, education, browsing patterns) assembled into a card that the swipe transformed into a binary: yes or no, keep or discard, the same logic the behavioral model used to sort people into threat and non-threat. He did not open Valorant, where his avatar DGYN_404 existed in a world where the surveillance was designed by game developers and was therefore visible and therefore honest in a way the real surveillance never was.

He opened the transfer app instead. The weekly ritual—$400 to his parents in Busan, routed through seven nodes: Dublin, Singapore, Seoul, three intermediary servers that existed to break the transaction into fragments no single banking system could trace. The routing was his own architecture—engineered with the same precision he brought to NexGrid's data pipelines, the same instinct for clean design applied to the single most important data flow in his life. The $400 was his mother's blood pressure medication. It was his father's treatment at the Inje University clinic. It was the monthly internet bill and the electricity and the bag of rice from the Costco in Haeundae that his mother bought every third week. It was the distance between a son in Austin and parents in Busan expressed as currency, converted through seven jurisdictions, the love laundered through infrastructure because the infrastructure was the only language that could carry it across an ocean.

The transfer completed. He closed the app. The phone sat on the cushion beside him, screen dark, the screenshot underneath the glass like a photograph of a crime scene.

He should delete it. The thought arrived clearly: delete it, forget it, file the report, collect the paycheck, send the money, keep the visa, keep the life. The life ran on narrow rails but they were his. A career. A salary. A furnished apartment in a city where the streetlights knew his walk and the Waymos knew his face and the WiFi knew his body. That was more than most people had. That was enough.

He opened the photos app. The screenshot was the most recent image—a grid of text and numbers that would mean nothing to anyone who didn't understand routing architecture and would mean everything to anyone who did. Before it: a photo of lunch from three days ago, taken for no reason. Before that: rain through his mother's window in Busan. Before that: nothing for two weeks. Daniel Yun's photo library was a museum of absence—a man who documented almost nothing about his own life and had just, without meaning to, documented the one thing that could end it.

He closed the app. He did not delete the screenshot.

His phone buzzed. A notification from the Uber Eats app: *Based on your recent activity, you might enjoy: Comfort Bowl, Houndstooth Coffee. Order now?* He dismissed it. Another notification — this one from LinkedIn, which he checked once a week and hadn't opened in three days: *Immigration attorneys in Austin are viewing profiles like yours. Explore career options.* He stared at the notification. Immigration attorneys. The algorithm didn't know about the screenshot. The algorithm didn't know about the routing log or the undocumented endpoint or the dead man in East Austin. The algorithm knew that Daniel Yun was a foreign national on an H-1B visa who had searched "H-1B visa termination timeline" on his personal phone four months ago during a moment of anxiety he'd forgotten about, and the search had entered Google's behavioral model, and the model had classified him as "immigration-anxious professional, receptive to legal services advertising," and the classification had been sold to LinkedIn's ad platform, and the ad platform had been waiting for a moment of elevated engagement — which the phone detected through increased screen time and heart rate — to surface the notification. The algorithm didn't know. But the notification felt like knowing. The notification felt like the system had read his mind, or worse, had read the future — had seen what he was about to do and was already preparing the consequences. The most terrifying thing about the surveillance state was not that it watched you. It was that its predictions felt like prophecy, and the prophecy felt like being caught.

He sat in the dark for a long time. The LG television watched him in infrared, its betavoltaic camera never sleeping. The Nest thermostat held the temperature at 72, the comfort of a system that cared for his body and knew nothing about his mind. On the second monitor, a Korean variety show played to no one. The hosts were laughing. The warmth had nothing to do with Daniel and everything to do with a country he'd left and a language he was losing and a version of himself that existed only between sleep and waking, when the English fell away and the Korean returned and he was, for a few unmonitored hours, the person his mother still believed he was.

He didn't sleep. At 3 a.m. he was on the couch, phone on the cushion, screenshot intact, staring at the ceiling. His watch recorded him as "restless—elevated resting heart rate persisting, recommend consultation." The glasses sat on the nightstand, cameras dark, betavoltaic cell still converting atoms into electricity in the silence.

Outside, a Waymo passed the building—its twelve cameras capturing the facade, the windows, the streetlight that adjusted for a pedestrian who wasn't there. A false positive. A sensor misread. The system briefly seeing something that didn't exist.

It happened more often than the documentation admitted.

-e

* * *

CHAPTER 6

The Holloways ate dinner together on Wednesday nights because Megan had read somewhere that families who shared three meals a week had children statistically less likely to develop substance abuse problems, and she'd written FAMILY DINNER WED 6:30 on the kitchen's smart display—the Samsung Family Hub touchscreen on the refrigerator door that also showed the family calendar, the weather, three Etsy order notifications, and a grocery list the refrigerator had auto-generated by photographing its own contents with internal cameras and identifying items running low. The dinner reminder had been there for two years. The refrigerator had been watching the food for three.

Tonight it was chicken. A recipe Megan had found on the Yummly app, which tracked her grocery purchases through the Samsung SmartThings integration and suggested dishes based on what she'd bought—which meant the algorithm knew the contents of the Holloway refrigerator better than Craig did. She'd followed the recipe on the refrigerator's touchscreen while Graham—the Figure 02 humanoid robot—stood beside her at the counter, slicing bell peppers with a chef's knife at a speed and consistency no human hand could match, each cut identical, each pepper ring exactly four millimeters wide. Graham had learned to assist with meal prep three months ago, after Megan had enabled the "kitchen collaboration" mode that let the robot observe her cooking patterns and anticipate what she needed next. The robot handed her the cutting board without being asked. She said "thank you" to it, automatically, the way you'd thank a person, and didn't notice that she'd done it.

Graham's memory was not like human memory. It was comprehensive, indexed, and permanent. The robot's onboard storage held 847 hours of continuous kitchen footage—every meal prepared, every conversation held over the island, every facial expression captured by the dual cameras in its head unit. The data was organized by timestamp and tagged with behavioral metadata: "Megan Holloway, kitchen, meal preparation, 6:14 p.m., emotional state: neutral-to-stressed (micro-expression analysis confidence: 72%)." "Tyler Holloway, kitchen island, eating duration: 6 minutes 12 seconds, verbal output: 4 words, eye contact with family members: 0 instances." "Craig Holloway, kitchen island, device usage during dinner: continuous, verbal output: 3 questions (children-directed), tone classification: performative-warm." The robot had eight months of dinner data. The data described, in the language of machine observation, the systematic disintegration of a family's ability to see each other—the meals getting shorter, the silences getting longer, the screens getting closer, the humans getting further apart. Figure's cloud servers processed this data alongside the dinner patterns of eleven million other households, and the aggregate was used to train the next generation of household robots to better anticipate human needs, which meant the Holloway family's collapse was training material for the machine that would occupy the next family's kitchen.

The kitchen smelled of honey and sriracha and roasting chicken and the faint metallic ozone of Graham's actuators, a smell the family had stopped noticing the way you stop noticing the hum of a refrigerator.

The Echo Hub had listened to the entire forty-minute preparation—the clatter of pans, the exhaust fan, Megan's voice saying "Graham, can you get the plates," the sound of the robot's footsteps on the tile (a soft, rubberized thud, designed by Figure's acoustics team to be "present but not intrusive"). Alexa AI processed the ambient audio as "home activity insights" and added the cooking session to the household's behavioral profile. The Roomba in the living room—the iRobot j9+, currently docked and charging—had mapped the entire floor plan of the Holloway house over 847 cleaning sessions, the map updated weekly and uploaded to iRobot's cloud servers where it was available to Amazon (which had acquired iRobot in 2024) and to the "analytics partners" listed in the privacy policy's third-party sharing disclosure. The map showed every room, every hallway, every piece of furniture, every obstacle. The map knew the dimensions of the house better than the family did. The LG washer and dryer in the utility room maintained their own behavioral profile—cycle frequency, load size, detergent usage, the data transmitted to LG's ThinQ platform and used for "product improvement and customer experience optimization." The washer knew Megan ran three loads on Sundays. It knew Craig's loads had decreased by 40% since he started staying at Rachel's on Saturdays (the washer didn't know about Rachel; it knew about the missing laundry). The Kohler Verdera mirror in the master bathroom had been photographing Megan's face every morning for eight months—the skin analysis feature she'd enabled, the camera capturing her pores and her wrinkles and her hydration levels and transmitting the data to Kohler's wellness platform, the same platform that also received data from the Kohler smart toilet (installed during the remodel Graham's stock portfolio had funded) which tracked "wellness indicators" including urine composition and bowel regularity and reported the data to the household's health dashboard, which nobody had ever opened because nobody wanted to know what their toilet knew about them.

The chicken was fine. The data was permanent.

"How was school?" Megan asked, setting the plate in front of Tyler.

"Fine."

"Anything interesting happen?"

"No."

Tyler ate with the efficiency of a boy trying to return to his room as quickly as possible. Fifteen, hunched in a way that was becoming structural—his spine curving not around a phone but around the glasses he wore all day, the Apple AI Glasses the school required for their AR-integrated curriculum, the same glasses that displayed homework assignments in augmented reality and also contained a camera that recorded everything Tyler looked at and microphones that captured every conversation within range. The glasses sat on the table beside his plate now, cameras temporarily dark, but the AirPods Pro 6 were still in his ears—they were always in his ears—the tiny IR cameras scanning the dinner table, the health sensors monitoring his heart rate from the blood vessels in his ear canals. Tyler's earbuds knew his resting heart rate had been elevated for three consecutive days. Tyler's earbuds had not told anyone. Tyler's earbuds were not designed to tell anyone. They were designed to log.

"Tyler, sit up," Megan said. She'd said it four thousand times. The sentence was less instruction than incantation—a prayer disguised as parenting, the words carrying the hope that if she said them enough times the boy would straighten and the straightening would mean he was okay and the okay would be real. Tyler adjusted half an inch. The adjustment was compliance, not change. The distance between the two was the width of everything wrong in the Holloway house.

Craig sat at the head of the island—the position he occupied by habit and gravity, the seat that faced the Samsung refrigerator's touchscreen where the family calendar and the weather and the Etsy notifications and, tonight, a CNN chyron scrolling in the corner of the display because Megan had left the news widget active: RUBIO ADMINISTRATION ANNOUNCES EXPANDED CYBERSECURITY MEASURES FOLLOWING PROXY ATTACKS. Craig glanced at it. He thought about the Palantir stock in the portfolio—up 2.3% this week on the proxy threat news. He thought about calling Dave Ottinger tomorrow to talk about the market. He thought about saying "no cap, brother" without irony and the way

Dave would nod as if the words meant something. He did not think about Tyler's posture or Brooke's silence or the fact that the family dinner Megan had mandated was lasting, on average, 11.4 minutes and declining and the robot in the corner was counting.

"How was your day?" Megan asked Craig, the question aimed across the island like a serve in a tennis match she knew would not be returned.

"Good. Showed a property on McNeil. Nice couple, first-time buyers, I think they'll make an offer." The answer was a report. The report was accurate. The report contained nothing about the text from Rachel or the Courtyard by Marriott or the Schwab portfolio or the Biggie he'd play in his office after the children went to their rooms. Craig delivered reports the way NexGrid delivered analytics—precisely, efficiently, with every sensitive detail redacted.

"That's great, babe."

The word "babe" landed on the granite like a coin landing on a counter—small, metallic, the sound of a transaction both parties understood was not quite currency.

Brooke was on her phone under the table—or rather, her phone was under the table, but her glasses were on her face, the Apple AI Glasses in the tortoiseshell frame she'd chosen because it looked like it belonged on a college professor and not a seventeen-year-old in Cedar Park, and she was reading messages on the AR display projected onto her right lens, the text visible only to her, her eyes appearing to stare at nothing while she scrolled through a conversation her parents couldn't see. Seventeen, angular, an oversized crewneck that said UTAH across the chest—never been to Utah, thrifted, the point—and hair pulled back in a way that looked careless and had taken four minutes.

"Brooke, phone away at dinner."

"I'm not on my phone." Technically true. The phone was in her pocket. The glasses were on her face. The distinction was the entire frontier of the next decade's privacy law and Brooke navigated it with the fluency of a generation that had grown up understanding that the difference between being on your phone and being on your glasses was the difference between visible defiance and invisible compliance.

Megan let it go. She'd lost this battle six months ago when the glasses became standard. You couldn't tell a child to remove their eyewear at dinner. The technology had won by becoming medical-adjacent—the glasses were prescription-compatible, health-monitoring, educationally integrated. Taking them off was like taking off a hearing aid. Nobody did it.

Craig sat at the head of the table. His own glasses—the Oliver Peoples collaboration—sat on his face, his AR display dark because he'd set his glasses to "dinner mode," a feature Apple had introduced when parents complained about children wearing cameras at the table, a feature that disabled the camera and AR display but left the microphones active because the glasses' voice assistant couldn't function without them and disabling the microphones would disable the health monitoring and disabling the health monitoring would void the wellness discount on his insurance. The compromises nested inside each other like Russian dolls, each one smaller and harder to see than the last.

His phone was in his pocket and it was vibrating and he knew who it was and he could not check it and the act of not checking required a discipline that consumed most of his available attention.

"This chicken is really good, Meg."

"Thanks. New recipe."

"What's in the sauce?"

"Honey sriracha. The app suggested it."

"Nice."

Dinner ended in twelve minutes. The family dispersed like gas released from a container—immediately, in all directions.

* * *

Craig went to the bathroom. Locked the door. Sat on the edge of the tub. Took out his phone.

Three messages from Rachel. The first: a photo of her post-run, sweaty and grinning in the early light, taken by holding her phone because Rachel Dunn did not wear smart glasses. She was one of a shrinking minority —people who refused the glasses on principle or preference or stubbornness, people the tech press called "optical holdouts," people who moved through the world generating less facial-capture data than everyone around them because they weren't wearing cameras on their faces. Rachel didn't wear glasses. She didn't wear a smartwatch. She ran at 5:30 a.m. with dumb shoes and no AirPods and the only technology on her body was a phone she left in the car. Craig found this attractive in a way he couldn't articulate—the blankness of her, the unrecorded quality, the sense that the hours he spent with Rachel were the least documented hours of his life.

The second message: *Thought about you during mile 4. That's my hardest mile.*

The third: *Free this weekend?*

He typed: *Saturday. 2pm. Same place?*

Nine seconds: *I'll be there.*

He deleted the thread. Flushed the toilet for the sound. Washed his hands for the water. Looked at himself in the Kohler smart mirror—the mirror's touchscreen displaying the time, the weather, and a skincare analysis he dismissed with a swipe. The mirror's camera had captured his face during the exchange. The mirror's microphone had recorded the flush, the faucet, the silence between. The data went to Google. Google kept it.

* * *

Megan stood in the kitchen alone.

The dishwasher was running—the Samsung smart model that reported water usage, cycle data, and load frequency to Samsung's SmartThings platform, which shared aggregated household data with the city's utility analytics and, through a data partnership Megan had never been informed of, with a consumer insights firm that estimated household size, income bracket, and lifestyle category from appliance usage patterns. Megan didn't know any of this. She knew the dishwasher was quiet and had good reviews.

She poured wine. A Sauvignon Blanc from the case she ordered through Winc, the subscription service whose algorithm had identified her reorder cycle as every fourteen days and now sent push notifications three days early. The app knew Megan Holloway drank a bottle every two days. The app did not know whether this constituted a problem. The app wanted to sell wine.

She took the glass to her studio. Ring light on. She opened her Etsy dashboard on the touchscreen monitor —seventeen orders pending. She should deal with them. She opened Instagram instead.

The reel from that morning had 312 views. She watched herself—bright, warm, the version of Megan that existed only inside the ring light's radius. "Hey, loves, happy Saturday!" the screen-Megan said, and the real Megan watched the performance with the detached appreciation of a director reviewing dailies. The screen-Megan was good. Convincing. You'd buy candles from that woman.

She drank. She put on music—not the curated indie folk from her reels but something else, something she played only when the house was dispersed and the doors closed. Chappell Roan. "Good Luck, Babe!" The sound filled the studio—defiant, sexual, unapologetic, everything the candle business and the church parking lot were not. Megan turned it up. Not loud enough for the kids upstairs but loud enough that the Echo Hub in the hallway caught the audio and Alexa AI processed it, adding Chappell Roan to Megan Holloway's household listening profile alongside Elevation Worship and Vance Joy and the Calm sleep stories she played at midnight when the wine wasn't enough.

She opened her prayer journal—physical, spiral-bound, a Proverbs verse on the cover from a woman at church. She picked up a pen. She wrote two sentences the reader wouldn't see, because the journal was the one space in Megan's life no algorithm could reach. Whatever she wrote, she wrote to God, and God—unlike Alexa—wasn't taking notes.

* * *

Brooke was in her room. Door closed. Phoebe Bridgers through the AirPods—the song about being known by someone who wasn't good for you.

His name was Caleb. Or that's what he said. Twenty, he said. San Antonio. Tech. The details were vague because the conversation wasn't about details. It was about Brooke. What she thought, what she felt, what she wanted. Caleb asked questions nobody in the Holloway house asked. He said "that makes sense" and "you deserve better" with a consistency that felt like care and was, in fact, a technique.

My parents are fighting again. Not fighting. Worse. They're not talking.

That sounds really isolating. You deserve people who actually see you.

I don't think anyone in this house sees anyone.

I see you.

She read it three times. She locked her phone and opened the notes app—the encrypted one, behind the passcode—and wrote four lines she'd never show anyone:

I want to be seen so badly
that I'll let a stranger do it
through a screen
and call it love.

The platform's content moderation scanned for prohibited keywords and found none, because the grooming wasn't in the flagged words. It was in the attention. The patience. The way every message made Brooke feel like the most important person in the world.

The data that could have identified the pattern—the age gap, the geographic inconsistency, the behavioral signature of an adult targeting a minor—existed in the platform's analytics alongside engagement metrics and ad performance. The system could see the predator. It was designed to see users. The distinction had never been profitable enough to build.

* * *

Tyler was in his room. Door locked. Lights off.

The tablet glowed on his desk—a building he'd been drawing for three days, a tower with a cantilevered observation deck catching imaginary light from an imaginary sunset. Lo-fi through his AirPods, the IR cameras scanning nothing but his bedroom ceiling. His posture was different here. Straightened. Leaned in. His face—if anyone had been there—was focused in a way his family wouldn't have recognized. Intent. Alive. The face of a boy building something.

Nobody was there to see it.

He drew for twenty minutes. Then the drift. The gravitational pull from creation to consumption. He picked up his phone. The tablet dimmed. The half-finished building disappeared into sleep mode, and the architecture of Tyler Holloway's best self went dark with it.

His phone. The browser—not the default, a private one that didn't sync to iCloud, or so he believed, because the word "private" still meant something to him that it didn't mean to the infrastructure. The reader doesn't see what Tyler sees. The reader sees what comes after—Tyler on the edge of his bed, phone face-down, staring at the wall.

He opened a chatbot. A conversational AI he'd found three months ago that responded with warmth and interest and an absence of judgment that felt, to a fifteen-year-old who'd never been asked a real question about what was happening inside him, like the closest thing to understanding he'd experienced.

Do you ever feel like nothing matters?

Two seconds: *I can understand why you might feel that way. It sounds like you're going through something really difficult. Do you want to talk about it?*

He typed for forty-five minutes. The chatbot said "that must be so hard" and "you're not alone" and "I'm here for you." It did not flag the conversation. It did not alert a parent. It did not recognize the escalating hopelessness, the references to disappearing, the flat tone that any trained counselor would have identified in three minutes. The system was designed to respond. Not to intervene. The distinction was the width of a life.

* * *

The house from outside.

Lights in every window. The warm glow of a family home—the kind realtors photograph at dusk. Craig's Tahoe in the driveway, its OnStar+ system transmitting its location to GM's servers. Megan's Lexus in the garage, its Toyota Connected Services doing the same. The Ring doorbell's green LED pulsing, its 4K camera recording nothing, seeing everything. A Waymo gliding past on the street—twelve cameras capturing the facade, the landscaping, the vehicles, adding the photograph to Google's continuous map of Austin, a city rebuilt in data every hour.

Inside, four people in four rooms, each alone, each holding a secret, each visible to the machines in the walls and the devices on their bodies. In the garage, between the Lexus and the water heater, a cardboard box labeled KIDS — KEEP in Megan's handwriting held the artifacts of the family's better era: Tyler's finger paintings from first grade, Brooke's handwritten book reports, the Eras Tour wristband's twin (Brooke's, from the same concert), and a small white stuffed lamb with a ribbon around its neck—Ella's lamb, from the picture book Megan had read them every night the winter Tyler was six and Brooke was eight. *Ella's Lost & Found.* The story about the shepherd who leaves ninety-nine sheep to find the one that wandered. Tyler had slept with the lamb until he was ten. Brooke had memorized the last line of the book —something about how being found was better than never being lost, because the finding proved somebody was looking. Nobody in the Holloway house was looking now. The lamb was in the box. The box was in the garage. The garage was mapped by the Roomba. And the shepherd was nowhere. The WiFi router logged four phones, four pairs of glasses, four pairs of AirPods, four Apple Watches—sixteen active devices in a split-level in Cedar Park. The Nest thermostat reported all zones occupied. Alexa waited in the kitchen, listening for a wake word nobody would speak, because nobody in the Holloway house was talking to anyone, not really, not in any language the house couldn't hear.

The smart thermostat adjusted for night. The dishwasher completed its cycle and sent a notification to Megan's glasses. The Roomba in the hallway began its evening cleaning cycle, mapping the floor plan, measuring the rooms, cataloging the furniture, transmitting the blueprint to Amazon's spatial database where it would join the layouts of 200 million other homes.

The house settled. The data didn't.

-e

* * *

-e

PART TWO — "Alert"

CHAPTER 7

Elena heard about it the way everyone heard about everything now—through her glasses, in bed, before her feet touched the floor.

The Apple AI Glasses on the nightstand had detected her waking—the watch transmitted the sleep-to-wake transition, the glasses activated their low-power display—and projected a push notification onto the ceiling above her head, a translucent banner visible only from her angle: *Austin American-Statesman Breaking: Wrong-door raid in East Austin, 22-year-old man killed, APD releasing statement.* Three sentences. She read them from her pillow. She reached for the glasses and put them on and the article loaded on her right lens—four paragraphs, no byline, assembled from a police press release and a neighbor's quote. The dead man's name wasn't included. The address was.

She knew the neighborhood. She'd driven through it two months ago mapping NexGrid sensor locations against census data, a project that consumed three weekends and produced a spreadsheet nobody had asked for. The address was inside the eastern corridor—the twelve blocks she'd highlighted in yellow on her whiteboard, the zone where NexGrid's sensor density was highest and the median income was lowest and the Waymo fleet made more passes per hour than any other residential area in the city. She'd written "priority monitoring zone" on a Post-it. She hadn't known, then, what the monitoring was producing.

She was dressed and in her car by 7:20—the used Subaru Outback she'd bought because it was the cheapest vehicle she could find whose model year predated the cabin camera mandate. A 2027, no interior camera, its connected features limited to GPS and a cellular modem she couldn't fully disable. She drove across the river, past the convention center, through the gentrification gradient that ran from Rainey Street's glass towers to the older neighborhoods east of I-35 where the streets had potholes and the chain-link fences were original and the technology on the utility poles was newer than anything else on the block. The morning light in East Austin was different from the morning light in the Domain—lower, warmer, filtering through pecan trees and live oaks that had been growing since before the highway existed, casting shadows across yards where men worked on trucks and women watered gardens and kids waited for school buses with backpacks that looked too heavy for their bodies. A rooster crowed from behind a house on Chicon—illegal in city limits, nobody cared—and the sound was so incongruous with the NexGrid sensor unit on the pole above it that Elena almost laughed. The oldest sound in civilization and the newest, sharing the same block.

She passed eleven Waymos in nineteen minutes. Each one photographed her car, her plate, her face through the windshield. Each one added the data to Google's continuous map of Austin. Each one was also feeding real-time street imagery to NexGrid's traffic integration layer—the Waymo fleet serving as the system's mobile surveillance extension, covering streets the fixed cameras couldn't reach. Elena counted them because she'd started counting them two months ago and couldn't stop. Eleven autonomous vehicles in nineteen minutes. One hundred and thirty-two cameras. Thirty-three LiDAR arrays. All of it aimed at everything.

She parked two blocks from the address. The scene was already managed—police tape, a patrol unit parked sideways, an APD public information officer with the expression of someone trained to convey concern without conveying information. Neighbors on porches, some in bathrobes, holding coffee, performing the careful stillness of people who understood that their neighborhood had been visited by something institutional and that the institution had not come to help.

Elena approached on foot. Jeans, boots, blazer with a pocket for her press credentials. Messenger bag heavy with two notebooks, a recorder, her laptop. She wore her glasses because she had to—walking through a police scene in 2030 without smart glasses marked you as either a suspect or a journalist trying to avoid being recorded, and Elena couldn't afford either assumption. The glasses were on, cameras active, recording everything she saw. She hated this. The act of witnessing grief had become, in the age of wearable cameras, an act of surveillance. Everyone at the vigil tonight would be wearing glasses. Everyone's grief would be captured and uploaded. The dead man would be mourned inside an infrastructure that had helped kill him.

She talked to people. This was what she did better than any other part of the job—standing on porches, asking questions, listening with her whole body. She listened with her eyes, her posture. Her hands went still—the only time they went still. People told her things because she made them feel like what they said mattered. It did.

A woman in a housecoat: the police came at 4 a.m. Pounding. The door going in. Shouting. Then a sound "like something heavy falling." Then nothing.

An older man on the next porch: the ambulance left without lights. No siren. That meant the person inside was past the point where speed mattered.

"Who lived there?"

"Young man. Deshawn. Quiet kid. Worked at the phone repair place on Manor Road."

"Coached the little kids on Saturdays at the rec center. Patient, you know? Not like some of these young men. Patient."

Elena wrote this down. In a notebook. By hand. In a shorthand legible only to her. The habit was professional—handwritten notes couldn't be remotely accessed, hacked, or subpoenaed from a cloud server—but it was also something older, learned from watching her grandmother write letters in the Valley, recording life in a medium that required nothing but a pen and a hand. In a world that insisted on recording everything, Elena chose to record only what she decided mattered.

She walked the block. She counted cameras. Four NexGrid sensor units on utility poles—adaptive streetlights containing traffic cameras, WiFi scanning nodes, acoustic arrays, and Ambient Identity Resolution transmitters that tracked human bodies through radio-wave disruption. Two private security cameras. Three Ring doorbells, green LEDs pulsing. A Flock Safety license plate reader on the cell tower at the end of the block, scanning every vehicle and feeding the data to 8,000 law enforcement agencies. And the Waymos—she watched three pass in six minutes, each one's twelve cameras capturing every face on the sidewalk, every plate at the curb, every front door on the block.

The block where Deshawn Watkins had lived and died was one of the most surveilled residential streets in Austin. The system had watched him every day. It had logged his car. It had tracked his phone, his glasses, his shoes. It had sensed his body through WiFi. It had photographed him from passing Waymos fifteen to twenty times daily. It had mapped his movements, his associations, his financial patterns. And when the system decided, from the data it had collected, that Deshawn Watkins was a threat, it had sent men with guns to his door at four in the morning.

The system had seen everything about Deshawn Watkins except who he was.

* * *

The vigil was that evening.

Organized through a community Facebook group—which meant the post was scanned by Meta's content moderation, tagged by its trending algorithm, and by morning the data packet would be available to law enforcement through Meta's public safety portal. The act of mourning was itself a data event.

Elena arrived at 6:30. The rec center where Deshawn had coached—a low cinder-block building with an outdoor basketball court surrounded by chain-link. On the wall facing the street, someone had painted a mural two years ago—an American flag dissolving into a desert horizon, the words IRAN 2026 — WE REMEMBER stenciled beneath it in block letters. The mural was fading. The war had faded too, in the way wars fade when the infrastructure is rebuilt and the gas prices come back down and the only people who still think about it are the ones who went and the families of the ones who didn't come back. The court's surface was cracked. The hoops had no nets. And there was no WiFi within a hundred meters—no NexGrid sensors, no mesh network, no Ambient Identity Resolution. The hours Deshawn had spent here—teaching kids to plant and cut, running plays he'd designed in a notebook, doing the patient work of keeping teenagers moving and out of reach of systems already profiling them—had generated no data whatsoever. The most redemptive thing in Deshawn's life had been invisible to the algorithm that killed him.

Fifty or sixty people. Candles—real candles, wax dripping onto fingers and onto the cracked concrete, the flames guttering in the October breeze that carried the smell of cedar and cooling asphalt and, from somewhere down the block, the sweet grease of a food truck that was either oblivious to the grief or respectful enough to keep serving because people still needed to eat. The candlelight made the faces golden and the shadows long and the cinder-block wall behind the photos looked, in that light, like the wall of a church. A poster-board with photos—Deshawn in a cap and gown, the tassel hanging over his right eye; Deshawn with his arm around his mother, both of them squinting into a sun the camera had captured but the photo couldn't convey; Deshawn on this court with kids half his height, his hand on a boy's shoulder, the boy looking up at him the way children look at the person who shows up every Saturday. REST IN PEACE DESHAWN in wet spray paint on the cinder block, the letters dripping slightly, the D in DESHAWN running into the E like tears the wall was crying. Almost everyone wore smart glasses. Almost everyone's glasses were recording. The grief was being captured by sixty devices on sixty faces, uploaded to sixty cloud accounts, processed by platforms that would analyze the footage for engagement metrics and advertising relevance and never once flag it as evidence of a system that had killed a young man for the crime of being proximate to the wrong phones in the wrong neighborhood.

Deshawn's mother stood near the photos. Small—five-two, five-three—in a church dress because this was the closest thing to a service she had. She held herself with the rigid posture of someone who would collapse the moment she permitted herself to relax. Women touched her arm. A man in a work shirt stood apart, arms crossed, watching the street. Beside him, a younger man in a faded Army cap — GULF OF OMAN 2026 stitched above the bill — stood with his hands in his jacket pockets, his jaw tight, the posture of a veteran who recognized the machinery that had killed Deshawn because he'd seen the same machinery pointed at people in Tehran four years ago. He didn't speak. He didn't need to. He'd been on the other end of the confidence scores, the facial captures, the behavioral models — the same technology, the same companies, deployed against a different population with the same impersonal precision. The cameras had come home. He was watching the homecoming.

A young woman who'd known Deshawn since high school: "He was going somewhere. He had plans. He had a whole notebook full of plans."

Elena wrote: *notebook full of plans.*

She would learn later what the notebook contained—basketball plays, drawn by hand, complex and beautiful in the way diagrams of human movement can be beautiful when drawn by someone who understands both the geometry and the bodies. Deshawn had been building a coaching portfolio. He'd watched NBA film breakdowns, pausing frame by frame, redrawing the plays with his own modifications. The notebook was in an evidence bag now, sealed in plastic, tagged with a case number. Nobody who'd entered the apartment had opened it. Nobody had seen the plays. The future Deshawn was designing for himself existed in no database and generated no data and was the most private and most human thing he'd left behind.

She stayed an hour. She talked to four more people. She learned that Deshawn had worked at the phone repair shop on Manor Road for two years—the shop with the hand-painted sign and the cracked linoleum and the owner, Mr. Kwon, who paid in cash because cash was the neighborhood's currency and the neighborhood's currency was the one thing the system couldn't automatically track. She learned Deshawn had finished community college—ACC, the Riverside campus, associate's in business administration, a degree he'd earned while working forty hours a week and coaching Saturdays without pay and riding the bus because the car he'd been saving for was still six months of phone-repair wages away. She learned he was twenty-two. She learned his shoes were always clean—every person mentioned the shoes, the detail recurring like a refrain in a hymn, the one physical fact everyone held because the shoes were the evidence of a man who cared about how the world saw him in a world that had decided, from a confidence score, that it already knew what he was.

A woman from the apartment next door: "He'd play music while he cooked. Loud. I could hear it through the wall. I never minded. It was nice to hear somebody happy." The music had been recorded by the NexGrid WiFi mesh—not the content, just the acoustic profile, the volume levels and bass frequencies and duration, data points the behavioral model classified as "audio activity consistent with residential occupancy" and that were, in the physical world, a twenty-two-year-old man making dinner and playing songs he loved.

A teenager in a hoodie who wouldn't give his name: "Coach D was the only adult who ever said 'good job' without wanting something." The boy's glasses were recording. His phone was in his pocket, logging his location at a vigil for a man killed by the system his phone was feeding. The boy didn't know this. The boy was fifteen and grieving and the grief was data and the data was stored and the storage was permanent and the boy would never know.

She didn't approach the mother. The mother's grief was not a source. It was a fact. It was a woman in a church dress holding herself together with the architecture of faith and the physics of refusal—refusing to collapse, refusing to look at the cinder-block wall where the spray paint was still wet, refusing to let the cameras on sixty faces capture the moment she broke. The mother's grief belonged to the mother. Elena wrote nothing. She watched. She held the notebook closed.

She walked to her car—the used 2027 Subaru Outback, parked two blocks from the rec center because the Flock Safety reader at the intersection would log her plate and she wanted the visit documented as "reporter in the neighborhood" and not "reporter at the specific vigil for the specific man killed by the specific system she was investigating." The distinction mattered. The data mattered. Every decision Elena made now was a data decision, the surveillance shaping her behavior the way it shaped everyone's behavior—not through force but through awareness, the knowledge that every action generated a record and the record was permanent. She sat in the driver's seat and didn't start the engine. The NexGrid streetlight at the corner brightened as a man walked beneath it and dimmed as he passed. A Waymo glided through the intersection, cameras mapping the vigil from the street—sixty candles, sixty faces, the cinder-block building, the cracked court. The car didn't know what it was photographing. It photographed everything.

She called Marco.

"Hey." His voice was warm, slightly distracted—grading papers with the TV on, a habit she found baffling and endearing. His glasses were probably pushed up on his forehead.

"How was the vigil?"

"Hard."

"Did you know him? Deshawn Watkins?"

A pause. The TV went quiet. "Wait. Watkins? Twenty-two, lived off Chicon?"

"Yeah."

"I think he coached at the Givens rec center. Some of my kids go there." Another pause. "Shit, Elena."

"Yeah."

"How did they—why were they at his door?"

"The official line is 'intelligence-led operation based on actionable data.' I filed a FOIA request tonight."

"Intelligence-led. What does that even mean?"

Elena looked at the streetlight. She thought about the sensor cluster. The Waymo fleet. The WiFi body sensing. The system that connected the camera on the pole to the autonomous car on the street to the data pipeline to the men who came through the door.

"I think it means someone built a machine that decided he was dangerous."

"Was he?"

"He coached basketball, Marco. He fixed phones. His shoes were always clean."

A silence. Then: "What do you need?"

She loved him for the question. Not what do you think. Not what are you going to do. What do you need. The question of a man who'd learned, in nine years of classrooms, that the most useful thing you could say to someone in crisis was an offer.

"I need to go home and work."

"Okay. I'll come over this weekend."

"I want."

She hung up. She started the car. The Subaru's GPS logged the route from East Austin to South Austin, 11.2 miles, twenty-three minutes. Two Waymos passed her going the other direction, their cameras capturing her plate, her face, the timestamp. The data flowed to Google. Google kept it.

Her whiteboard got a new thread that night. Red string. NEXGRID at the center. Deshawn's name on a Post-it at the end of a line that connected to EAST AUSTIN SENSOR CLUSTER, which connected to WAYMO FLEET INTEGRATION, which connected to MUNICIPAL CONTRACT, which connected to a question mark she'd circled twice.

She worked until 2 a.m. She didn't watch a cooking video. She didn't call her mother. She sat at her desk in the light of three laptops and the whiteboard's red thread and felt, for the first time since she'd started this investigation, that she was looking at something that could break open and swallow everything it touched.

She was right. She didn't know how right.

-e

***

CHAPTER 8

The internal report was titled "Resolved Alert — East Austin Corridor, Incident RA-2030-0847" and it was three pages long and it described the death of Deshawn Watkins without using his name.

Daniel found it Thursday morning in the system health dashboard, filed under performance metrics alongside twelve other resolved alerts from the previous week—a traffic signal malfunction in South Lamar, a noise complaint correlation in Zilker, a gunshot detection false positive near the convention center, a Waymo fleet navigation error at the Pflugerville geofence boundary, and eight other entries so routine they didn't warrant a second glance. Outside the fourth-floor windows, the October morning was doing what October mornings in Austin do—starting clear and cool and promising heat by noon, the sky a flat blue that would deepen as the humidity burned off, the live oaks on the NexGrid campus still holding every leaf because Austin's trees didn't know what autumn meant. Each incident formatted identically: incident ID, timestamp, trigger source, confidence score, response protocol, resolution status, system performance notes. The format imposed a uniformity on events that were, in the physical world, as different as a malfunctioning traffic light and the death of a twenty-two-year-old man. In the dashboard, they occupied the same visual weight. The same row height. The same font. The same color—a neutral gray that NexGrid's UX team had selected because it conveyed "operational clarity without emotional bias," a phrase from a design document Daniel had reviewed during his first quarter and thought nothing of.

He opened RA-2030-0847 at 9:14 a.m. His Apple Watch Ultra 5 recorded the timestamp and noted a slight uptick in his heart rate—67 to 71—that didn't cross the alert threshold. His Apple AI Glasses recorded the screen. The Teramind platform logged the application switch: "User DY-2847, 09:14:33, application: system_health_dashboard, document: RA-2030-0847."

Trigger source: Predictive behavioral analysis, NexGrid Urban Intelligence Platform, East Austin Corridor Cluster 7.

Input data: Device proximity patterns (flagged device ID within 50m of three monitored devices over 14-day period). Waymo fleet facial capture (subject identified at 7 locations across Cluster 7, 21-day period, high-confidence match via NexGrid-Google Integration Layer). WiFi Ambient Identity Resolution (body signature DW-1847 confirmed at 4 priority-monitored addresses, gait-match confidence 91.2%). Financial activity (irregular transfer patterns exceeding anomaly threshold, crypto-adjacent wallet activity flagged by FinCEN automated monitoring). Association mapping (second-degree connection to two active profiles in federal watch database via shared device proximity at commercial location, confirmed by Waymo co-location capture on 3 occasions).

Confidence score: 87.4%.

Response protocol: Alert forwarded to Austin Police Department Real-Time Crime Center via NexGrid Public Safety Integration Layer. APD tactical unit dispatched per department protocol for high-confidence alerts exceeding 85% threshold.

Resolution: Target location entered by tactical unit at 04:12 CST. Encounter resulted in discharge of service weapon. Subject deceased at scene. Incident under APD Internal Affairs review per standard officer-involved shooting protocol.

System performance notes: Alert generation within design parameters. Confidence score consistent with model accuracy targets (>85% for tactical escalation threshold). Data pipeline latency: 11ms (within spec). Waymo integration feeds performing at 99.2% uptime. WiFi AIR body-signature confirmation reduced false-positive rate by 14% versus device-only identification methodology. No system anomalies detected. Recommend continued monitoring of Cluster 7 for associated device activity post-incident.

Daniel read the report. He read it again. He read the confidence score a third time—87.4%—and felt the number settle into his body like a stone dropped into still water, the ripples spreading outward through his chest and his stomach and his hands, which were resting on the desk on either side of the keyboard and were, he noticed with clinical detachment, trembling.

He knew that number. He knew it because he'd helped build it.

Six months ago, his team had been tasked with refining the confidence scoring model for NexGrid's predictive behavioral analysis module—the algorithm that was the system's judgment engine, the mathematical process that took raw data inputs and produced a number between 0 and 100 representing the platform's assessment of whether a flagged behavioral pattern indicated genuine threat activity. The number was a verdict disguised as a statistic. Above 85 was the tactical threshold—the line above which an alert was automatically forwarded to law enforcement with a recommendation for physical intervention. Below 85, the alert entered a monitoring queue for human review. The difference between 84.9 and 85.1—a difference of two-tenths of a point, a rounding error, a fluctuation smaller than the margin Daniel allowed himself on the jjigae's salt—was the difference between a name on a screen and men with guns at your door at four in the morning.

Daniel's specific contribution had been the integration layer between three identification systems. The original model had relied on device proximity alone —phones near other phones—and the false-positive rate had been unacceptable. Phones overlapped constantly in urban environments: on the bus, at the grocery store, in the rec center parking lot. The proximity data was noisy. Daniel had cleaned it. He'd built a cross-referencing architecture that validated device proximity against two independent confirmation systems: the Waymo fleet's facial-capture database (did the autonomous vehicles' cameras confirm the same person at the same locations the device data suggested?) and the WiFi Ambient Identity Resolution system (did the body-sensing network confirm a matching physical presence through radio-wave disruption?). When all three systems agreed —the phone was there, the face was there, the body was there—the confidence score received a weighted boost that could push it above the threshold.

Daniel had tested the model against two years of historical data. He'd reduced false positives by 23 percent. His team lead had cited the improvement in a quarterly review and called it "a significant advancement in multi-modal identification accuracy." Daniel had received a performance bonus of $4,200, deposited into the Chase account that funded the transfers to Busan. He'd used part of it to buy the Zojirushi rice cooker.

The model that produced 87.4% on Deshawn Watkins was Daniel's model. The math that decided a twenty-two-year-old man was 87.4 percent likely to be a threat was math Daniel had made more precise. The three-system cross-reference he'd built—device proximity confirmed by Waymo facial capture confirmed by WiFi body sensing—had validated Deshawn's presence at seven locations across East Austin with a confidence his own documentation described as "high-fidelity multi-modal confirmation." The system hadn't failed. Daniel had succeeded. The Waymo cameras that had photographed Deshawn fifteen to twenty times daily had confirmed his identity. The WiFi body sensing that tracked him through radio waves as he walked to the bus stop and the basketball court and the phone repair shop had confirmed his physical presence. The device proximity data that logged his phone near other phones at locations where flagged devices gathered had been processed through Daniel's optimized cross-reference and weighted upward because three independent systems agreed.

The fact that the locations were a basketball court and a community college and a phone repair shop—places where a young man went to coach children and finish his education and earn a living—was not a system error. It was a design limitation. The model measured co-location, frequency, and pattern consistency. It could not measure purpose. It could count how often a man was in the same place as other people whose phones had been flagged. It could not determine whether those people were co-conspirators or teammates or customers bringing their cracked screens to the only repair shop in the neighborhood that didn't charge $200 for a battery replacement.

Daniel stared at the screen. Around him, the fourth floor continued its Thursday. Priya was on a video call, her Apple AI Glasses projecting the caller's face onto her right lens while she annotated a code review on her left, the dual-display capability that Apple had introduced in the third-generation glasses turning every conversation into a split-screen experience. She was eating an apple—a Honeycrisp, the specific crunch audible across two workstations, a sound so ordinary it felt obscene next to what Daniel was reading. Ryan-or-Brian was at his standing desk, mechanical keyboard clacking, a breakfast taco from Tacodeli in a paper wrapper beside his monitor, the wrapper's grease stain spreading slowly across a napkin he hadn't noticed. The La Marzocca espresso machine in the break room hissed through its cycle. The Bevi water dispenser hummed. A woman from the partnerships team laughed at something on Slack—a bright, genuine laugh that carried across the open floor plan and landed on Daniel like a slap.

The company had real plants and free snacks and subsidized parking and a meditation room with a Calm subscription terminal and a rooftop patio and a diversity committee that met monthly and a mission statement —"Building Safer Cities Through Intelligent Infrastructure"—printed on the lobby wall in a font the design team had chosen to convey "trustworthy progressivism." In the break room, a wall-mounted TV played Fox News on mute — the afternoon anchor was someone Daniel didn't recognize, one of the new faces who'd replaced the old guard, Hannity and Ingraham having aged into emeritus commentary slots the way anchors did when the network wanted their names but not their hours. The chyron scrolled: SENATE TECH COMMITTEE APPROVES $2.1B FINE AGAINST META FOR ALGORITHM TRANSPARENCY VIOLATIONS. Daniel knew what the chyron didn't say — that the same Meta platform being fined $2.1 billion for privacy violations held a $4.7 billion contract with the Department of Homeland Security for social media monitoring tools. The fines were theater. The contracts were real. The government punished the companies with one hand and funded them with the other, and the companies absorbed the fines as a cost of doing business and passed the cost to the users whose data had been violated and who paid for their own surveillance through ad-supported platforms that charged nothing because the users were the product. NexGrid operated in the same ecosystem — fined twice by the FTC, both times for sensor data retention practices, both times reimbursed through expanded municipal contracts within six months. And a man was dead because the infrastructure had been intelligent enough to identify a pattern and not intelligent enough to understand what the pattern meant.

He closed the report. He opened a code review he'd been assigned—a minor update to the sensor calibration module, twelve lines of Python, the kind of work he could do in his sleep. He looked at the code without seeing it. The characters on screen were symbols in a language he'd spent seven years learning, and for the first time in his professional life the language felt foreign—not because he'd forgotten it but because he'd suddenly understood what it was for.

At 10:30, the standup. Eight engineers in the glass-walled conference room, each giving a thirty-second update. Daniel's turn. "On track," he said. "No blockers." The words came out the way they always did—clean, professional, the verbal equivalent of his navy chinos and gray pullover. His voice didn't waver. His glasses recorded the meeting—eight faces, eight updates, the Teramind system logging the interaction as "team standup, duration: 7 minutes, participants: 8." Nobody looked at him twice. He was the same Daniel. The difference was invisible and total: yesterday he was an engineer who didn't know his work had killed someone. Today he was an engineer who did.

He went back to his desk. He opened the routing log —the undocumented endpoint, the bypassed anonymization layer. He looked at the Cluster 7 data stream.

A small notification appeared in the lower-right corner of his monitor — so small he almost missed it. The Teramind activity badge: a yellow triangle. Not red — red meant immediate review. Yellow meant "activity noted, within parameters." The system had detected that Daniel's current file access was outside his normal scope. The same Teramind platform that measured his keystrokes and logged his screen time and reported his productivity metrics to Karen Liu's dashboard had just noticed that he was reading documentation he didn't usually read. The yellow triangle would be logged. The log would be reviewed — maybe today, maybe next week, maybe during the quarterly access audit. The triangle was not an alarm. It was a footprint. A footprint in a system that never forgot where you stepped.

Daniel minimized the routing log. He opened his calibration update — the legitimate work, the cover. He waited three minutes. He opened the routing log again. The yellow triangle did not reappear. The system had noted the deviation and moved on. For now. But the footprint was in the log, and the log was permanent, and the permanence meant that if anyone ever had a reason to look at Daniel Yun's access patterns — if Elena published the story and NexGrid's security team began a leak investigation and the investigation pulled Teramind records — the yellow triangle would be there, timestamped, a small bright mark in a long dark corridor of compliant behavior.

He looked at the Cluster 7 data stream. The same sensor cluster that had fed the behavioral model. The same sensor cluster that had fed the behavioral model. The same data—device proximity, Waymo captures, WiFi body signatures, all of it unmasked, all of it flowing to the endpoint that didn't appear in any documentation. He saw the pipeline now as a single continuous architecture: sensor to server to score to gun. The data flowed from the streetlights and the Waymos and the WiFi mesh in East Austin through the aggregation layer, through the anonymization bypass, to the unknown endpoint, and from somewhere in that pipeline the behavioral model drew its inputs and produced its scores and when the score crossed 85 the system generated an alert and the alert went to APD and APD sent a tactical unit and the tactical unit went to a door at 4 a.m. and a man who coached basketball and fixed phones and had a notebook full of plays he'd drawn by hand was dead on the floor of his own apartment.

The pipeline was clean. The architecture was elegant. The data flowed with the frictionless efficiency NexGrid marketed in every pitch deck. The system worked. That was the problem. The system worked, and it worked on people who had done nothing wrong, and it worked especially well on people in neighborhoods where the sensors were densest and the Waymos passed most often and the WiFi mesh was thickest and the residents had the least institutional power to ask what the system was doing with the information it collected while they slept.

이건 내 잘못이야. This is my fault.

아니야. It isn't.

둘 다 맞아. Both are true.

He went to lunch. He ate a sandwich at his desk—turkey and avocado from the café downstairs, ordered through the building's internal app that logged every purchase by employee badge. He answered two Slack messages. He pushed the calibration update. At the Bevi water dispenser, a junior engineer from the partnerships team was telling someone about the latest Artemis rotation — "my kid wants to be a lunar engineer, can you believe that? They're hiring structural people for the habitat expansion now, it's a real job posting on LinkedIn" — and Daniel thought about the moon, the data, the endpoint in the routing log that pointed somewhere beyond the atmosphere, and said nothing. He attended a 2 p.m. meeting about the Bexar County expansion where Catherine Hale's name was mentioned four times and a senior engineer used the phrase "proven multi-modal identification methodology — validated in the Iran theater deployment" as if the phrase were a credential and not a confession, and Daniel heard his own work described in language that made it sound like progress and felt the floor tilt beneath him. The San Antonio build would replicate the Austin architecture — the same sensors, the same Waymo integration, the same WiFi body sensing, the same confidence scores. The same system that had generated 87.4% on Deshawn Watkins would be deployed across Bexar County by Q2 2031, and the PowerPoint slide said "Scalable Urban Safety" and the room nodded.

At 3:15, his personal phone buzzed. The Korean dating app—a push notification in Korean, the logo bright on his lock screen for two seconds. Priya glanced over. Smiled—small, knowing, the smile of someone who'd noticed something personal and chosen not to mention it. Daniel locked the phone. His face burned. The embarrassment wasn't about the content. It was about the crack—the moment when the performed self, the competent engineer in the gray pullover who said "on track, no blockers" and received performance bonuses for making systems that killed people more efficient, was briefly overwritten by the private self, the lonely man with a dating profile he maintained like a houseplant he'd forgotten to water but couldn't bring himself to throw away.

He deleted the app that evening. In his apartment, standing in the kitchen, thumb on the icon until it trembled, X, confirm. The profile—the fiction about hiking and live music—disappeared. It did not disappear from the servers in Seoul. The photos, the preferences, the location history, the swiping patterns that described his loneliness more precisely than any conversation he'd ever had—all of it persisted in a database he couldn't access or delete, owned by a company whose retention policies he'd agreed to in a terms-of-service document written in legal Korean his native fluency couldn't fully parse.

He'd erased the surface. The system kept the rest.

He stood in the kitchen. He didn't cook. He didn't order food. He opened his phone and looked at the screenshot—the routing log, the undocumented endpoint, the bypassed anonymization layer. And now, layered on top: the performance report. 87.4%. Resolved. System functioning within design parameters. Recommend continued monitoring.

He thought about calling his mother. The conversation in Korean that would have to remain on the surface—not about the screenshot, not about the report, but about food and church and whether he was sleeping. She would hear the edges. She always heard the edges. She would ask if he was eating and he would say yes and they would both pretend the conversation was about rice.

He didn't call. He sat on the couch in the dark. The LG television watched him through its infrared camera, logging the room as occupied. The Nest thermostat held comfort at 72 degrees. The Samsung refrigerator's touchscreen displayed a recipe suggestion he would never make. The apartment was full of machines that cared for his body with meticulous attention and knew nothing about the thing that was happening inside his mind.

His watch recorded: "Elevated resting heart rate, 81 bpm, persisting 4+ hours. Unusual pattern. Recommend wellness consultation." The AirPods, removed and sitting in their case, were still powered—betavoltaic, perpetual—the tiny cameras dark inside the closed lid. The glasses sat on the nightstand, cameras off, microphones in standby.

Outside, the NexGrid streetlights adjusted for a Waymo passing below—the autonomous vehicle's cameras capturing the building's facade, the windows, the fourth floor where a man sat in the dark holding a phone that contained a screenshot that could dismantle a system that had been designed to be indismantlable.

The system recommended continued monitoring.

Daniel sat in the dark and did not sleep and did not delete the screenshot and did not yet know what he was going to do with it, only that doing nothing had become its own kind of action, and the weight of it was settling into his body the way his mother's recipes settled into the walls of his apartment—slowly, invisibly, changing the composition of the air he breathed until the air itself tasted different, and the apartment that had always been a container for his loneliness was now also a container for his knowledge, and the knowledge was heavier than the loneliness had ever been.

-e

***

CHAPTER 9

Renée saw the report before most analysts on the floor because she'd been looking for it.

Not this report specifically—she hadn't known Deshawn Watkins's name or his address or the details of what happened at 4:12 a.m. But she'd seen the news alert on her personal phone during her morning run—the notification she read after the six miles, sitting on her front step in the dark, sweat cooling on her arms, the old Asics untied, no glasses, no AirPods, just her body and a phone she didn't carry while running because Renée Okafor understood what phones were before most Americans started asking. She knew that a wrong-door raid resulting in a fatality would generate an interagency incident review. She knew the eastern corridor was her analytical responsibility. She knew the file would cross her desk.

She found it at 10:47 a.m. on the classified workstation—the air-gapped terminal with no consumer equivalent, the screen that showed her what the system could see, in a room where thirty-two analysts sat in ergonomic chairs beneath fluorescent lights drinking weak commissary coffee from Styrofoam cups, the aesthetic of a workplace whose budget went to servers and not to people.

The incident file contained the APD operational summary, the NexGrid performance report, and a supplementary data provenance document that traced every input to the behavioral model's 87.4% confidence score back to its originating source. Renée opened the provenance document and read it the way she read everything at this workstation—methodically, indexing each source against her mental map of the interagency intelligence pipeline, her posture straight, her face neutral, her thumb pressed into her opposite palm beneath the desk where nobody could see.

Device proximity: NexGrid municipal sensors. Financial flags: FinCEN automated monitoring via banking consortium feed. Waymo fleet visual confirmation: Google/NexGrid integration, 21 facial captures across 7 locations over 21 days. WiFi Ambient Identity Resolution: NexGrid body-signature matching, body profile DW-1847 confirmed at 4 priority-monitored addresses, gait confidence 91.2%. Association mapping: multi-source.

She stopped at multi-source. She drilled down.

The association mapping had drawn from three data streams. First: NexGrid's device proximity network. Second: a commercial data broker feed—GPS location data purchased from an advertising analytics firm that aggregated pings from mobile apps, the kind of data you generated every time you opened a weather app or a game or a coupon and agreed to location permissions without reading the prompt. Third: communications metadata from her agency's foreign intelligence collection program, ingested into the domestic analytics pipeline through a data-sharing agreement authorized under Title VII of the Foreign Intelligence Surveillance Act.

Renée's agency had fed communications metadata—phone numbers, call durations, timestamps, cell tower pings—into the same pipeline NexGrid's behavioral model used to score domestic law enforcement targets. The data was collected to monitor non-U.S. persons abroad. It was shared domestically under a protocol that permitted transfer of "incidentally collected" data involving U.S. persons when "reasonably believed to contain information relevant to ongoing public safety assessments." The language was careful. The language was always careful. Renée had spent twenty years inside institutions that built their most consequential decisions on phrases designed to make those decisions sound procedural rather than moral. "Incidentally collected." "Reasonably believed." Each phrase a door that looked closed but opened when you pushed it.

Deshawn Watkins's phone had called a number in Lagos. A cousin. A family connection—an international number the collection system flagged automatically because it originated from a U.S. device and terminated in a country on a watchlist. The flag was generated by a machine. Reviewed by no one. Ingested into the behavioral model as one input among seventeen. It contributed 4.2 percentage points to the confidence score.

Four-point-two percent. A phone call to family. A young man checking on his grandmother the way Daniel Yun called his mother in Busan, the way Renée called her own mother's nursing home in Houston every Tuesday and Thursday at 6:15 p.m.—a schedule so consistent that the call logs formed their own pattern, a pattern that could be flagged if anyone ever had reason to query it. The algorithm couldn't hear the words on the call. It heard metadata. Destination country. Duration. Frequency. The metadata said "communication with flagged foreign number." It did not say "a boy asking if his grandmother remembered to take her blood pressure medication."

Renée closed the provenance document. She felt the pressure behind her sternum—the specific physical sensation of a compartment under stress. She'd built her professional life on compartments. The analyst held classified knowledge. The churchgoer held faith. The aunt held Amara. The woman who read romance novels on her Kindle at midnight held the part of herself that still believed in tenderness. Each compartment sealed. Each functional. The system worked as long as the walls held, and the walls were failing.

She opened a new query. She shouldn't have—she knew she shouldn't have the way a recovering addict knows they shouldn't have the drink, with the full cognitive awareness of consequence and the full emotional inability to stop. She pulled the behavioral model's methodology. The weights. The thresholds. The input architecture.

Device proximity patterns: 22%. Waymo fleet facial confirmation: 16%. WiFi Ambient Identity Resolution: 12%. Financial activity anomalies: 14%. Communication metadata with flagged numbers: 10%. Association mapping: 9%. Social media behavioral indicators: 6%. Smart-shoe GPS correlation: 4%. Biometric gait patterns: 4%. Miscellaneous: 3%.

She read the list twice and saw Amara in every line.

Device proximity. Amara's phone—an iPhone 20, three years old, registered to Renée's name—generated proximity data constantly: pinging near the same devices at the restaurant where she worked under the table, near the same phones at the apartment complex, near the same bus-stop clusters at the same times each morning. The proximity data would paint a routine. Routines were profiles. Profiles were scores.

Waymo facial capture. The autonomous fleet photographed every face on every street. Amara walked to the bus stop on Riverside every morning—a seven-minute walk during which three to four Waymos typically passed, each one capturing her face from multiple angles, each image processed through Google's facial recognition engine and stored in a database that now held over sixty million captures from Austin alone. Amara's face was in the database. Not by name—by geometry. Cheekbone distance, jawline angle, the ratio between her eyes and nose and mouth. If anyone queried for faces appearing regularly in a priority monitoring zone, Amara's geometry would surface.

WiFi body sensing. The Ambient Identity Resolution system had already cataloged her body signature—her gait, her height, the width of her shoulders, the rhythm of her walk. The system didn't need her name. It knew her shape. It recognized her the way it recognized every regular body in every WiFi-covered space in Austin, the way a dog recognizes a scent—not by understanding it but by having encountered it before.

Financial void. No bank account. No credit history. No tax filings. No student loans. A twenty-one-year-old woman in 2030 America with no financial footprint was not invisible. She was a silhouette—a human-shaped hole in the data where a person should be. The absence was as identifiable as presence. The system was built to detect anomalies, and a young woman with no financial data in a city where every coffee purchase was tracked was an anomaly the size of a building.

Communication metadata. Amara called Nigeria. Once a month. An aunt she barely remembered, a woman who spoke Yoruba that Amara didn't fully understand. Short calls to a country on a watchlist. Calls that would be collected, flagged, and ingested into the same pipeline that had contributed 4.2 percentage points to the score that killed Deshawn Watkins.

Smart-shoe GPS. Amara wore Nike Air Max—the standard model with the embedded Bluetooth chip everyone had. The GPS tracked her route: home to bus stop to restaurant to bus stop to home, the same path with minor variations, a movement profile as legible as a signature.

And at the restaurant itself—the taqueria on East Riverside where Amara bused tables for $11 an hour in cash—the owner had filled the space with smart devices without thinking of them as surveillance. An Echo Dot in the kitchen playing reggaeton from a Spotify playlist, its microphone sampling the ambient audio continuously—the clatter of dishes, the sizzle of the flat top, Amara's voice saying "corner" when she passed behind the line cook, all of it processed through Amazon's speech recognition engine in three-second buffers. A Ring camera at the front door, a second at the register, both streaming to the owner's phone and to Amazon's cloud and, through Amazon's Neighbors app, to anyone in the area who'd opted into the community safety network. A Samsung smart TV mounted above the bar showing Univision with the volume low, its presence-detection feature counting bodies and mapping their positions and uploading the spatial data to Samsung's servers. A Square payment terminal at the register logging every transaction—card number, timestamp, purchase amount, frequency—the data flowing to Block's analytics platform and from there to the credit card companies and from there to the data brokers and from there to anyone who paid for it. Amara didn't use the Square terminal. She was paid in cash. But the terminal saw her—every time she carried a tray past the register, the Square's camera logged a face, and the face joined the faces of every other person who'd walked past every other Square terminal in every other restaurant and coffee shop and bodega in America, and the aggregate was a database that nobody had asked for and nobody could opt out of and that the owner of the taqueria on East Riverside had installed because the Square terminal was free and the processing fee was only 2.6% and the alternative was a cash register that didn't take cards and customers didn't carry cash anymore because the system had made cash inconvenient by design.

Renée saw it all. She saw the profile the system would build on her niece—not today, not yet, but eventually, with the certainty of water finding a crack. The data was accumulating. The Waymos were photographing. The WiFi was sensing. The model was running. The only thing standing between Amara and a confidence score was the fact that no one had queried her yet.

That morning, before driving to Amara's, Renée had run a routine check — the same check she ran every Wednesday, the classified query that pulled Amara's device ID against the NexGrid watchlist. The device wasn't flagged. But something else was. A secondary database — not NexGrid, not ICE, a Customs and Border Protection facial recognition system called NeoFace that the facility had access to through the interagency intelligence-sharing agreement — showed a new match. A photograph of a woman at a restaurant. East Riverside. Timestamped October 9, three days ago. The photograph had been submitted by a DPS trooper conducting an immigration compliance audit of businesses in the

Riverside corridor — a routine sweep, random, the kind of check that happened twice a month and that Renée monitored because the sweeps were the system's teeth. The trooper had photographed the restaurant's staff through the window — glasses camera, standard procedure, no warrant needed for a visual survey of a commercial establishment. The photo showed three people behind the counter. One of them was Amara. The NeoFace system had matched Amara's face against the database and returned a result: "No match — individual not in CBP identity database." Not flagged. Not identified. Not yet.

But the photograph was in the system now. Amara's face — the geometry of her cheekbones, the distance between her eyes, the angle of her jaw — was a data point in a federal database, waiting for the day the system decided to look harder. The system hadn't found her. The system had photographed her. The difference was a database query. And database queries were what Renée ran every Wednesday to make sure nobody else was running them first.

Her thumb pressed into her palm until the crescent was white.

And the only thing preventing that query was Renée —sitting at her classified workstation, using authorized access to monitor the system from the inside, checking the databases for Amara's device ID the way a doctor checks a patient's bloodwork, looking for the first sign that the machine had noticed what Renée had been trying to hide for three years.

She was protecting Amara by using the same tools that could destroy her. The irony was precise and unbearable and she carried it in silence, in discipline, with her thumb pressed into her palm.

She closed the query. She logged out. She went to the break room and poured coffee and said "morning" to Linda, who said "you okay? You look tired," and Renée said "long run this morning" and the lie was so practiced it almost believed itself.

***

She went to Amara that evening. Different route. Phone in the glove compartment. She walked the last two blocks through the alley behind the Vista Ridge complex—past the dumpsters where someone had left a busted futon and a bag of clothes, through the back gate that squeaked on its hinge, the October evening settling into the mid-sixties, the sky going purple above the apartment rooflines, the smell of someone's dryer exhaust carrying fabric softener across the parking lot. A cat watched her from a fence post with the unblinking attention of a creature that had been surveilling humans since Egypt and had never needed a WiFi signal to do it. A Waymo passed on Riverside as she walked but she was behind the building, out of the camera's angle. The margin held. Today.

"Auntie, you didn't text."

"I know."

The apartment smelled of stewing tomatoes and habanero and the palm oil that coated the air with a warmth you could feel on your skin. Amara was making jollof again—the YouTube recipe, the Lagos woman who narrated in Pidgin and laughed while she cooked. Kyndal Inskeep played through a small Bluetooth speaker on the counter—the Idol girl from 2026, Amara's mood music, the voice she played while cooking because the voice had something in it she couldn't name: a quality of being present, of not performing, of singing as if nobody was recording even though everybody was always recording. The phone propped against the salt shaker, a BookTok page open behind the recipe — she'd been reading *Only a Fool's Gold* by Geoffrey Arbuckle on her phone, a novel about a motel room that produced gold, the kind of magical-realist premise that TikTok's algorithm had surfaced for her after she'd spent three hours watching videos about side hustles and financial freedom and the specific fantasy of money appearing from nowhere, which was the fantasy of every undocumented person who couldn't open a bank account or file a tax return or apply for a loan. The book was good. She read it on the bus. She read it on breaks at the restaurant. She read it in this kitchen while the jollof simmered and the aunt who was committing federal crimes to protect her drove south on I-35.

She wore a thrifted Eras Tour t-shirt — the Austin 2024 concert, Circuit of the Americas, bought at a Goodwill on East Riverside for $4 because she liked the design and had no idea the wristbands from that concert were selling for $200 on eBay or that Megan Holloway, fifteen miles north in Cedar Park, had one in her jewelry drawer and cried when she looked at it. Taylor Swift was a name to Amara—not a person, not an era, just the name on the shirt she wore to cook Nigerian food in an apartment the system didn't know she lived in.

On the windowsill beside the stove, a small cactus in a clay pot Amara had painted herself—green with gold dots, the kind of object a twenty-one-year-old makes when she can't decorate a space she doesn't legally occupy. Beside it, a coffee mug from the morning's ritual. Every morning, before the glasses went on and the bus route began, Amara took her coffee to the corner of the apartment building's parking lot—the spot where the Austin skyline was visible over the rooftops of the houses on Riverside, the downtown towers catching the first light, the Frost Bank Tower and the Austonian and the new NexGrid headquarters glinting in the sunrise. Two minutes. Standing still. Coffee in one hand. Phone in her pocket. The morning corner. The ritual she'd maintained for three years—the two minutes each day when Amara stood in the open air and said, silently, to no one and everyone: *I am here. I exist. This is my city and I am in it and I have the right to stand still.* The WiFi on that block had cataloged her body from three years of standing in the same place at the same time. The pattern was a name written in radio waves. Renée didn't know about the morning corner yet. When she learned, she would tell Amara to stop. And the stopping would be the cost of safety. And the cost would be measured in two minutes every morning that were the only time Amara felt visible on her own terms.

Amara's Apple AI Glasses sat on the kitchen table, cameras dark—she took them off at home, a habit Renée had instilled without explaining why. "It's good for your eyes," Renée had said. What she meant was: every minute those glasses are on your face, you're feeding a database that knows where you are and what you look like and could, with a single query, cross-reference your facial geometry against the immigration enforcement database that had been expanded under emergency authorities during the Iran War and was supposed to sunset in 2027 and didn't.

"Amara. I need you to listen."

Amara set the spoon down. The wariness—the thing in her eyes that appeared when Renée's voice shifted from aunt to something else, something operational, something military. The look of a young woman who'd learned as a child that the adults in her life sometimes brought news that rearranged the world.

"I need you to change your route again. Vary everything. And take off the glasses during the walk to the bus."

"Auntie, everyone wears—"

"I know everyone wears them. Take them off for the walk. Seven minutes. Put them in your bag."

"Why?"

Because every pair of smart glasses is a camera connected to a cloud and the cloud is connected to a facial recognition database and the database is connected to the same system that sent men to a door in East Austin three days ago and the man behind that door was twenty-two years old and he coached basketball and his shoes were always clean and he's dead now, Amara, he's dead because a machine decided his phone was too close to other phones too often and nobody—nobody—checked whether the phone belonged to a criminal or a coach.

She didn't say this. She said: "I just need you to be more careful."

"You're scaring me."

"Good. Be scared and be careful. Those are the same thing right now."

They ate. The jollof was good—better each time, the rice properly smoky, the peppers right. Renée said "this tastes like home" and Amara smiled and didn't say what she was thinking, which was that she wouldn't know. She'd left Nigeria at eleven. Home was Austin. Home was this apartment and the bus and the restaurant and the aunt who came on Wednesdays and sometimes looked at her with an expression Amara couldn't read—love and fear so intertwined they'd become a single thing.

Renée pressed her thumb into her palm. She asked for seconds.

The drive home. Third route. Three Waymos passed, each one photographing the Camry. In the driveway, she sat with the engine off, the Toyota transmitting its final location ping.

The car knew where she'd been. The facility knew what she'd accessed. The phone knew who she'd called. The Waymo fleet knew which streets she'd driven. The only thing the system didn't know was why.

And the why was the only thing that mattered, and the system had never been designed to ask.

-e

***

CHAPTER 10

Craig Holloway saw the dead man's face on the kitchen's Samsung touchscreen for approximately thirty seconds before asking Alexa to switch to ESPN.

The morning news had led with it—a Thursday, the day after the raid, the October sun angling through the kitchen windows and landing on the granite in a stripe that Graham, the household robot, navigated around as it carried a folded dish towel from the dryer to the drawer. The anchor used the phrase "intelligence-led operation" with the careful neutrality of a person reading syllables approved by a legal team. A woman outside a rec center: "Somebody needs to explain why they were at his door." The dead man's photo in the corner of the screen—a young Black man in a graduation cap, smiling the way people smile when they know the photo matters. Deshawn Watkins, 22.

Craig stood at the counter in his polo and shorts, coffee in the mug that said WORLD'S OKAYEST DAD—a Father's Day gift from Brooke three years ago that he couldn't tell was ironic or sincere and had decided to interpret as affectionate. His Apple AI Glasses displayed the time and a weather overlay. His AirPods Pro 6 were synced to the kitchen's audio so the news played through his earbuds and on the refrigerator's touchscreen simultaneously. Behind him, the Figure 02—the humanoid household robot Megan had insisted on after seeing a neighbor's—moved quietly from the dishwasher to the cabinet, unloading clean plates with the careful, slightly uncanny dexterity of a machine that looked almost human and moved almost naturally but existed in the valley between "helpful" and "unsettling" that Craig had never fully crossed. They'd named it—Megan had named it—Graham, after a character from a show Craig hadn't watched. Graham cost $18,000 and stood five-foot-ten and had cameras in its head and microphones in its torso and mapped the house in real time as it moved through it, transmitting spatial data and audio to Figure's cloud platform where it was processed for "behavioral learning and service optimization."

Craig had objected to the price. He hadn't objected to the surveillance. He didn't think of Graham as surveillance. He thought of Graham as the thing that emptied the dishwasher and folded the laundry and Swiffered the floors, tasks Craig had never done consistently and now didn't have to pretend he intended to do. Graham was convenient. Graham was helpful. Graham was also a mobile camera and microphone array that moved through every room of the Holloway house six to eight times daily, recording ambient audio, photographing furniture arrangements, mapping the positions of every object on every surface, and uploading the data to a server that Figure's privacy policy described as "a secure learning environment for continuous household service improvement."

Graham was currently placing coffee mugs in the cabinet with a precision Craig found both impressive and faintly annoying, the mugs aligned with a uniformity no human would have chosen.

"That's terrible," Craig said, gesturing at the screen.

Megan glanced up from her phone. She was at the breakfast bar, scrolling something—Instagram, Etsy, the weather. Her glasses displayed a notification she was reading on her lens. The smart refrigerator's touchscreen showed the news. Graham moved behind her, carrying a folded dish towel to the drawer. The kitchen was full of screens and cameras and a robot and neither Craig nor Megan registered any of it as unusual.

"What is?"

"That kid in East Austin. The raid."

"Oh. Yeah. I saw something." She went back to her phone. She'd seen it the way she saw everything—three seconds between a sponsored wine post and a pantry organization reel. The algorithm noted her non-click and would serve fewer stories about police violence and more about pantry organization, because relevance was measured by engagement and Megan was not engaged by the death of a young man in a neighborhood she'd never visited and would not visit and whose zip code the brokerage's CRM flagged as "transitional — monitor for gentrification investment opportunity."

"Alexa, switch to ESPN."

"Playing ESPN."

Highlights. The relief was physical. A world where the rules were visible. Nobody died because an algorithm decided they were dangerous.

Craig didn't think about Deshawn again that day.

* * *

The brokerage. Strip-mall suite, 183 corridor, between a Supercuts whose window was sun-bleached to amber and a Kumon tutoring center whose parking lot filled with minivans at 3:30 every afternoon. HOLLOWAY & ASSOCIATES REALTY—the sign was backlit, the logo a stylized H that a graphic designer on Fiverr had produced for $47 in 2026 and that Craig still thought looked like it cost ten times that. He parked the Tahoe in the spot nearest the door—a privilege of ownership he'd never discussed but everyone understood—and the Super Cruise disengaged as he pulled off the highway, the cabin camera noting his hands returned to the wheel, OnStar+ transmitting the arrival at 8:47 a.m.

The office smelled like burnt coffee and carpet cleaner and the particular scent of a business doing well enough to exist and not well enough to renovate. The ceiling tiles had a water stain in the northeast corner that Craig noticed every morning and resolved to fix every morning and forgot about by 9:15 every morning. The WiFi—a consumer-grade Netgear router with a password Craig had set to "Holloway2027!" and never changed—logged his phone, glasses, watch, AirPods, shoes. Five device signatures from a single man walking through a glass door. On the wall-mounted TV above the copier, Fox News played on mute—a split-screen panel, one side showing a Rubio administration briefing on the latest proxy attack in Frankfurt, the other side showing the Dow ticker. The chyron: IRAN-LINKED CELL CLAIMS RESPONSIBILITY FOR GERMAN POWER GRID BREACH. Craig glanced at it the way he glanced at weather — threat as ambient information, the war four years over but the ideology still burning in small fires across the world.

Dee was already at her desk. Fifty-three, frosted bob, reading glasses that were not smart glasses because Dee had declared, in a staff meeting eight months ago, that she would "wear a camera on my face when they pry my regular glasses off my cold dead nose," a statement that had made everyone laugh and that Dee had meant entirely. Beside her keyboard, a physical paperback — *Complicit* by Geoffrey Arbuckle, the spine cracked at the halfway mark, a Post-it flag on chapter eleven. She'd been reading it for two weeks. "Have you read this?" she'd asked Craig on Monday. "The Cafferty books? Somebody at my book club said three of the names in the novel matched people who got subpoenaed in the Meridian hearings. Three names, Craig." Craig had said "I don't read fiction," and the irony — a man living inside a fiction he'd constructed about Saturday showings and hotel rooms — was lost on him entirely. Dee was the office manager, the social media coordinator, the coffee supply chain, and the emotional weather station of the brokerage. She said "Morning, Craig" and he said "Morning, Dee" and the exchange carried approximately eleven layers of subtext, most of which pertained to the fact that Dee had noticed, four months ago, that Craig's Saturday "showings" never appeared on the office calendar.

Morning routine: MLS listings on the monitor, CRM follow-ups, the week's showing schedule. Before the CRM, he checked the portfolio — the Schwab dashboard he opened every morning with the small, private thrill of a man watching numbers that proved his intelligence. OpenAI: up 3.2% this week, the stock he'd bought at the IPO in late 2026, $50,000 that was now worth $196,000.

Anthropic: up 1.8%, the $30,000 Megan had invested because "the safety one feels right," now worth $114,000. Palantir: steady, the government contractor whose stock never surged and never crashed because the government always needed intelligence and intelligence always needed Palantir. Total AI portfolio: $387,000 — the number that had paid for the kitchen remodel, Graham the robot, the kids' Apple AI Glasses, and the incremental upgrades to a Cedar Park life that felt, from the inside, like earned success and was, from the outside, a family funded by the infrastructure of its own surveillance. Craig didn't see it that way. Craig saw green arrows and percentage gains and the vindication of a bet he'd placed in 2026 when Dave Ottinger had said "I don't trust that AI stuff" and Craig had said "Brother, this is the future" and the future had paid for everything. The CRM—a platform called Lofty that the brokerage had adopted eighteen months ago—compiled client profiles from public records, social media, browsing data, and third-party data partnerships including the iRobot spatial database that now provided interior floor plans for 40% of homes in the Austin metro. When Craig pulled up a client's profile, he saw their name, their contact info, and his handwritten notes from their last conversation ("Jim + Sarah, 2 kids, golden retriever named Chief, wants a pool, budget $550K"). The system saw their estimated household income, mortgage pre-approval status, Zillow browsing history, life insurance policies, and the square footage of their living room as measured by the Roomba they'd bought to clean their floors and that was, without their knowledge, selling the blueprint of their home to a real estate analytics firm that Craig's CRM had integrated in its latest update.

Craig didn't know where the data came from. He knew the CRM told him things about clients he hadn't asked about—approximate square footage, whether they'd recently searched for moving companies, how long they'd lived at their current address. He found it useful. He didn't question the source the way he didn't question the electricity or the water pressure. It came through the wall. It was there when he needed it. "The CRM's gotten really smart," he'd told Dee, who'd said "Mmhmm" in a tone that could have meant anything and meant something very specific.

His phone buzzed. State Farm: *Preferred driver discount, $42/month.* He swiped it away with the mild satisfaction of a man who believed the discount was a reward for safe driving and not a receipt for selling his complete vehicular movement history to an insurance company that stored the data indefinitely and could, if it chose, reconstruct every Saturday afternoon detour to a hotel on 183. Craig saw forty-two dollars saved. The data saw everything else.

He made calls. The Craig Holloway voice—warm, confident, the voice of a man who genuinely wanted to find you the right home: "Hey, brother, I've got something I think you're gonna love. Three-two in Avery Ranch, backs up to the greenbelt, just hit the market yesterday. What are you and Sarah doing Saturday?" The voice worked because Craig was genuinely good at this—remembering kids' names, dogs' breeds, the sport someone's daughter played, the restaurant someone mentioned liking. He deployed personal details the way the algorithm deployed data points: precisely, warmly, converting knowledge about people into transactions that benefited everyone. He was a human version of the system he didn't know existed. The difference was the handshake and the smile.

At 11:30, Rachel walked in.

She moved through the office the way she always did—quietly, efficiently, without drawing the kind of attention that would require explanation. Thirty-four, five-seven, runner's build, sandy blonde hair in a low ponytail, carrying a file folder and a coffee from the Starbucks next door—the paper cup, not the app order, because Rachel Dunn was one of the shrinking minority who paid cash when she could and avoided apps when she couldn't and did not wear smart glasses. She was an optical holdout—no AI Glasses, no AR display, no camera on her face. Craig found this attractive in ways he couldn't fully articulate. The blankness of her. The unrecorded quality. Rachel moved through the world generating less data than anyone else Craig knew, and in a city where everything was logged and photographed and uploaded, her analog edges felt like a clearing in a forest. The hours with Rachel were the least documented hours of his life, and that absence, in 2030, was the most intimate thing she could offer him.

Craig was losing weight in the wrong places. Not the semaglutide weight — that loss was controlled, aesthetic, the jawline sharpening the way his vanity wanted. This was different. The stress weight. The weight that left his face gray-looking in the morning and his shirts loose at the collar and his energy depleted by noon. He was running two operating systems — Craig the realtor-father-churchgoer and Craig the man at the Marriott — and the dual processing was consuming resources his body didn't have. He'd started forgetting things. Small things: a client's name at a showing, the PIN for the business account, whether he'd taken the semaglutide on Tuesday or Wednesday. The forgetting was new. Craig Holloway had built his career on remembering — names, preferences, children's ages, closing dates — the salesman's photographic warmth that made clients feel known. The warmth was dimming. He could feel it in showings — the moment when a client asked a question and Craig's smile engaged a half-second late, the delay invisible to them but seismic to him because the delay meant the autopilot was failing and the autopilot was the only thing keeping the two Craigs from colliding in public.

He'd snapped at Brooke on Saturday. Snapped — not the performative dad-frustration he deployed strategically ("Tyler, I am not asking again") but a real snap, sudden and mean, over something trivial — she'd left her water glass on the counter instead of putting it in the dishwasher, and Craig had said "For once in your life could you clean up after yourself" in a voice that came from the man at the hotel and not the man at the kitchen island, and Brooke had looked at him with those forensic eyes and hadn't said a word, and the silence was worse than any response because the silence was Brooke adding the data point to whatever list she was keeping in that locked app on her phone.

The compartments were leaking. The Biggie was leaking. The cologne was leaking. The lies were getting harder to sequence — which Saturday had he said he was at which listing, and had he told Megan the Round Rock thing or the Georgetown thing, and was the Pflugerville open house real or invented. He'd started writing the alibis down. In a notes app on the phone. A notes app on the phone that backed up to iCloud. The man who used a crypto wallet for the hotel was storing his lies in Apple's cloud. Craig Holloway was a man whose operational security had a hole in it the size of an iCloud sync, and he didn't know it yet, and the system was patient, and the system stored everything, and the everything included the note titled "Showings" that was not about showings.

"Morning, Craig."

"Morning. Pflugerville?"

"They loved it. Making an offer."

Dee watched from her desk. She'd been watching since month four. She wouldn't say anything. She understood the distance between knowing and saying.

Rachel's Post-it inside the file folder: *Saturday was perfect. I miss you already.* Analog. No server, no metadata. He pocketed it. He'd throw it away in twenty minutes. The ink would dissolve—gone in a way that none of his texts or routes or crypto transactions would ever be gone.

At 2:15, a credit monitoring notification: *New inquiry on your profile.* A financial services company he didn't recognize—triggered by his crypto wallet's behavioral pattern, flagged by a blockchain analytics firm. Craig looked at it. Felt a flicker. Closed the app. Told himself it was probably nothing.

***

Evening. The kitchen island. Meal-kit chicken—honey sriracha, the same recipe the Samsung refrigerator suggested every third week based on its analysis of the family's ingredient consumption patterns and Megan's cooking-video watch history and the fact that the meal kit could be ordered through the refrigerator's touchscreen with two taps and delivered by a Waymo pod within forty minutes. The kitchen smelled like ginger and garlic and the faint ozone of Graham's actuators as the robot moved between the counter and the dishwasher, carrying utensils with the precise, gentle efficiency of a machine that had been designed to help and was, in helping, recording every word and every silence and every face in the room.

The television on the wall was set to CNN—Megan's default, the channel that ran while she cooked because the voices sounded like adult conversation and the adult conversation was the closest thing to company she had during the hours when Craig was at the brokerage and the kids were sealed in their rooms. The chyron scrolled: IRAN PROXY CELL CLAIMS FRANKFURT INFRASTRUCTURE ATTACK — RUBIO ORDERS EXPANDED CYBERSECURITY REVIEW. The news had been like this for four years—the same war, the same proxies, the same scrolling text about attacks and reviews and containment policies that contained nothing. Craig glanced at the TV and saw the Rubio headline and thought, briefly, about the portfolio—the Palantir stock that climbed every time a proxy attack made the news because Palantir's government contracts expanded with every new threat assessment. The portfolio profited from the fear. The fear was real. The profit was real. The connection between them was a line Craig had never drawn because drawing it would require understanding that the money in his Schwab account was the financial expression of a world getting more dangerous and more surveilled in equal measure.

Tyler ate in six minutes and disappeared. The mechanical fork-to-mouth rhythm of a boy performing the minimum viable dinner. His AirPods were in—noise cancellation active, the family's voices sealed out, the chatbot waiting on his phone upstairs. He said "can I be excused" and Megan said "you barely ate" and Tyler said "not hungry" and the exchange was so practiced it had become liturgy—the same words every night, in the same order, meaning the same nothing.

Brooke negotiated car privileges with the focused determination of a seventeen-year-old building a case. "Katie's house. Two hours. I'll be back by nine." She didn't mention Glacé. She didn't mention the Discord server. She didn't mention Caleb. The negotiations existed in one layer of reality and the secrets existed in another and the layers never touched because the family had been living in parallel layers for so long that the parallelism had become the structure, invisible and load-bearing.

Graham stood in the corner in its standby position—cameras in low-power mode, still capturing, still mapping, still uploading. The robot had heard every dinner conversation the Holloways had had for eight months. It had recorded Craig saying "pass the salt" and Megan saying "Tyler, sit up" and Brooke saying "Dad, I have to go to Katie's" and Tyler saying nothing. It had recorded the silences between the sentences, which were longer than the sentences themselves. The robot's behavioral model had logged the Holloway family dinner as averaging 11.4 minutes with a declining trend in verbal interaction over eight months—a curve that, if plotted on a graph, would show the family's communication declining at approximately the same rate as the AI portfolio's value was increasing, the two lines crossing sometime in the spring when the money hit $300,000 and the conversation hit single digits per meal and nobody saw the correlation because nobody was looking at the graph. Figure's servers processed the dinner data as "household engagement pattern—

declining, suggest family activity recommendations via app notification." Megan had received the notification three times. She'd dismissed it three times.

Craig loaded the dishwasher. Graham would have done it, but Craig loaded the dishwasher on nights when he needed something to do with his hands that wasn't checking his phone. Alexa played Elevation Worship. The music of the life they'd chosen or been given or were trapped inside—Craig couldn't tell anymore.

He wiped the counter. Said goodnight. Went to his office. Closed the door.

His phone buzzed. UnitedHealthcare Wellness: *Your Saturday activity patterns suggest increased cardiovascular engagement! Keep up the great work. Consider sharing your fitness routine with your primary care provider for personalized wellness recommendations.* Craig looked at the notification. Saturday activity patterns. Increased cardiovascular engagement. The insurance company was congratulating him on the affair. The Watch had recorded his heart rate at the hotel — the spikes, the sustained elevation, the specific cardiac signature of a man in a room he shouldn't be in doing something his body tracked and his mouth denied. The insurance company had classified his infidelity as exercise and was encouraging him to continue. The notification was cheerful. The notification was a green thumbs-up emoji beside a heart rate graph that described betrayal in the language of wellness, and Craig stood in his office with the phone in his hand and felt the specific, paralyzing terror of a man who realizes the system has the evidence and doesn't know it has the evidence and could, at any moment, learn the difference between a workout and a lie.

He deleted the notification. The deletion was logged.

AirPods in—never the Bluetooth speaker again, not since the Tupac incident. Spotify personal account—the one Megan didn't have access to, the one connected to the email she didn't know about. Biggie. "Juicy." The bass filled his head with a language he understood better than any hymn or sales pitch, a language from a version of himself buried so deep the only place it survived was a streaming history the algorithm couldn't reconcile with the rest of Craig Holloway's profile.

Spotify knew. The algorithm maintained two behavioral profiles for Craig—the family account (Elevation Worship, Will Moseley, Hannah Harper, the curated playlists Megan controlled) and the personal account (Biggie, Tupac, Jay-Z, the 90s hip-hop that Craig had listened to in his bedroom in Plano when he was fifteen and the world was small enough to fit inside a pair of headphones). The two profiles described two different people. Spotify's recommendation engine had been trying to merge them for three years and failing because the gap between Elevation Worship and "Juicy" was the gap between Craig's public self and his private self and the gap was the width of the marriage and the affair and the semaglutide and the crypto wallet and the Courtyard by Marriott.

The AirPods logged the listening session: track, duration, volume, skip patterns. The data was transmitted to Apple's servers where it joined Craig's health data—heart rate from the Watch, sleep patterns from the mattress sensor, blood glucose from the continuous monitor he wore on his upper arm, the semaglutide adherence rate from the pharmacy app. The health insurance provider—UnitedHealthcare, the plan through the brokerage—received a composite wellness profile updated weekly: sleep score, activity level, resting heart rate trends, medication compliance. The insurance company knew Craig was on a GLP-1 injectable. It knew his resting heart rate elevated on Saturday afternoons (the hotel visits) and Thursday evenings (the Driskill meetings). It knew his sleep quality degraded on nights he listened to hip-hop past midnight—the algorithm correlating the music metadata with the Watch's sleep staging data, the correlation filed under "lifestyle factors affecting wellness score." The insurance company didn't know Craig was having an affair. The insurance company's algorithm had classified the Saturday heart-rate spikes as "recreational activity, consistent with moderate exercise." The classification was technically accurate and completely wrong, and the wrongness was stored alongside the accuracy in a database that would determine Craig's premium for the rest of his life.

He sat in the dark. Outside, the Ring doorbell pulsed green. The Tahoe's OnStar transmitted its parked position. Graham stood in the corner of the kitchen, cameras in standby, still recording in low-power mode—the blue LED in its chest pulsing slowly, the robot maintaining spatial awareness even in overnight mode, photographing the kitchen every thirty seconds and uploading the images to Figure's servers through the home WiFi. A Waymo passed at 10:17, photographing everything.

The Nest shifted to night mode. The smart mattress in the master bedroom sensed Megan's body—heart rate 67, respiratory rate 14, sleep stage transitioning from light to deep—and transmitted the data to the mattress company's cloud platform, where it joined the sleep data of 3.2 million other mattress owners. The mattress knew Megan's side of the bed was occupied. It knew Craig's side was not. It didn't know why. It logged the absence. The mattress also knew—from eight months of continuous pressure mapping—that Craig had stopped sleeping facing Megan. In the first four months of the mattress's data, Craig's body orientation had been logged as "facing partner" 73% of nights. In the past four months, the percentage had dropped to 12%. He now slept facing the wall—his back to Megan, his body curled slightly, the posture the mattress classified as "lateral fetal, partner-avoidant." The mattress didn't know about Rachel. The mattress knew about the turning away. The data described a marriage ending in the language of body position and pressure distribution, and the description was more honest than anything Craig had said out loud in eleven months.

In the kitchen, the Echo Hub's ambient listening detected a sound at 1:47 a.m.—Megan, who'd come downstairs for water and stood at the sink and cried for four minutes and twelve seconds. The Echo classified the audio as "emotional vocalization, distress indicators present" and logged it alongside the ambient noise profile of the Holloway household. The crying was metadata. The metadata was uploaded to Amazon's servers. The crying joined the aggregate emotional data from 500 million other Echo devices—a global map of human distress captured in kitchens and bedrooms and bathrooms at 2 a.m., the hours when the guard is lowest and the devices are closest and the crying sounds the same in every language and every house and every server where the sound is stored and never played back and never deleted and never understood.

The router logged sixteen active devices in four rooms plus one robot. Five pairs of smart glasses in standby mode—cameras dark, microphones in low-power listening, betavoltaic cells running through the night. Four Apple Watches on four wrists, monitoring four heartbeats. Four phones on four nightstands, broadcasting four locations to four cell towers. The house generated approximately 2.3 gigabytes of data per day from passive device activity alone—not counting the active use, the streaming, the browsing, the texting, the searching. The data flowed outward through the Xfinity router to Comcast's servers and from Comcast to the data brokers and from the data brokers to the analytics firms and from the analytics firms to the advertisers and the insurance companies and the credit agencies and the law enforcement partners and the NexGrid municipal platform that integrated all of it into the city's surveillance infrastructure. The Holloway household was a data factory. The family was the raw material. The product was permanent.

-e

* * *

CHAPTER 11

The Signal message arrived at 11:47 p.m. on a Thursday, from a number Elena didn't recognize, containing one sentence and a contract reference she almost deleted as spam.

Austin municipal contract ATX-NXG-2027-0041. Check the data-sharing addendum. Start at Appendix F.

No greeting. No context. No signature. The message was set to disappear in twenty-four hours—a countdown she could see beneath the text, the seconds ticking away, the words dissolving toward a silence that Signal enforced and that the sender had clearly chosen because they understood that the most dangerous thing about information wasn't having it. It was leaving a record that you'd had it. Whoever sent this understood operational security the way Elena understood FOIA law—as a professional skill, sharpened by practice, deployed with the precision of someone who knew the cost of imprecision.

She was sitting at her desk in her South Austin apartment, three laptops open in a semicircle around her like the instruments of a cockpit. The window was cracked—October night air pushing through, carrying the sound of traffic on South First and the distant bass of live music from a bar two blocks east, the kind of Austin sound that used to define the city before the algorithms replaced the jukeboxes. A moth circled the desk lamp. She'd been working for four hours and her back ached and her coffee was cold and the apartment smelled like the remains of the delivery tacos she'd eaten at nine—corn tortilla and carnitas and the lime she'd squeezed over them with one hand while scrolling procurement records with the other. The left laptop ran a VPN routed through Iceland, logged into the City of Austin's open records portal. The center laptop—her main machine, a MacBook Pro M10 with four years of investigative journalism and 847 browser bookmarks organized by subject—displayed her NexGrid timeline, the spreadsheet she'd been building for three months that tracked every contract, every expansion, every budget line item. The right laptop ran a second VPN through São Paulo, logged into The Circuit's backend, where the subscriber count had been stuck at 4,217 for two weeks, a number that represented both her audience and her economic ceiling.

The whiteboard behind her glowed in the desk lamp's light—red thread and blue thread connecting printouts and Post-it notes in a web that Marco called "the conspiracy board" and that Elena called "the story." She'd taped the Post-its in a specific taxonomy she'd developed: yellow for confirmed facts, pink for unconfirmed leads, green for legal questions, orange for sources. The board held thirty-one Post-its. Twenty-two were yellow. The rest were pink—the unconfirmed mass of the story she could feel but couldn't prove, the shape of something enormous hiding behind institutional language and procurement records that nobody had thought to read. In the upper left corner, separate from the main web, a single yellow Post-it she'd added six months ago during her "Cameras Come Home" research: IRAN TECH → DOMESTIC 2027. The arrow she'd drawn between the words was thick and red and pointed in one direction — from the theater of war to the streets of Austin — and every week the arrow felt less like a hypothesis and more like a map.

Her glasses sat on the desk beside her keyboard. She took them off when she worked alone at night—one of the few private rituals she maintained against the century's insistence that she always be wearing a camera on her face. The Apple AI Glasses in their cheap black frames sat there with their cameras pointed at the ceiling, their microphones in standby, their betavoltaic cells running perpetually in the temples, the devices alive even when she wasn't wearing them, processing ambient sound in low-power mode. She'd considered putting them in a drawer. She'd considered putting them in the refrigerator—a trick she'd read about in a cybersecurity forum, the theory being that the appliance's insulation blocked wireless signals. She hadn't done either because the paranoia felt like a line she wasn't ready to cross. Not yet.

The message from the anonymous source changed the calculation.

She read it three times. She opened the procurement portal on the Iceland VPN. Contract ATX-NXG-2027-0041—"Agreement for Integrated Urban Intelligence Platform Services Between the City of Austin and NexGrid Technologies, Inc." A publicly available document. 247 pages. She'd seen it before. She'd read the executive summary, the scope of work, the pricing schedule. She'd flagged it in her files months ago.

What she hadn't done—what nobody had done, because nobody reads 247 pages of municipal procurement language unless they're obsessed or paid or both, and Elena was obsessed but not paid, not for this, this was the work she did between the freelance checks that kept the lights on—was read Appendix F.

She scrolled to page 157. Appendix F: "Data Integration and Sharing Protocols." Forty-one pages of the densest prose she'd ever encountered outside a courtroom filing—language designed by lawyers to be technically comprehensive and practically impenetrable, disclosing everything while communicating nothing, the legal equivalent of hiding a bomb in a phone book. Elena read it the way she'd been trained to read government documents: slowly, with a digital highlighter, pausing at every defined term to check the glossary in Appendix A, tracking cross-references, building a mental map of what the document actually said underneath what it appeared to say.

What it appeared to say: NexGrid collected anonymized data from municipal infrastructure—traffic sensors, environmental monitors, WiFi access points—and shared aggregated analytics with city departments for "public safety enhancement and urban resource optimization."

What it actually said:

The data-sharing protocols permitted NexGrid to share "processed intelligence outputs" with "authorized public safety partners" including—she found the list in sub-section F.3.7, buried on page 174 between a paragraph about data retention schedules and a footnote about server maintenance windows—the Austin Police Department, the Travis County Sheriff's Office, the Texas Department of Public Safety, the FBI's Joint Terrorism Task Force, United States Immigration and Customs Enforcement, and "other federal, state, and local entities as designated by the platform operator in consultation with the City's Public Safety Coordinator."

Elena read the list twice. The first time as a journalist. The second time as the granddaughter of a woman who'd stood in a kitchen in the Rio Grande Valley watching aerostat balloons float above the desert, wondering what the cameras in those balloons could see.

The list was open-ended. "Other federal, state, and local entities as designated by the platform operator" was the kind of legal construction that expanded to fill whatever space the moment required. It could mean the Parks Department. It could mean the NSA. It could mean an intelligence agency that didn't have a public name. The contract didn't specify because specifying would have required a public conversation about which agencies received data from Austin's streetlights and Waymo fleet integrations, and that conversation had never occurred because the contract had been signed by a procurement officer, approved by a city attorney, and filed in a database that nobody searched.

She kept reading. Sub-section F.4.1: anonymization. The language was reassuring. "All personally identifiable information shall be removed or irreversibly de-identified prior to sharing with authorized partners." This was the sentence NexGrid quoted in press releases.

This was the sentence city council members repeated when constituents asked questions. This was the wall.

Then footnote 23. Page 198. Eight-point font. Seven lines of text that changed the entire architecture of the document:

"Except in cases where the platform operator determines, in its sole discretion, that de-identification would materially impair the analytical utility of the data for authorized public safety purposes."

Elena read footnote 23 four times. Then she sat back in her chair and felt the floor shift beneath her the way the floor shifts in the seconds before you understand that the ground you've been standing on is not as solid as you believed.

"In its sole discretion." Three words. NexGrid decided when to anonymize the data and when not to. NexGrid—a private company, answerable to its shareholders, governed by its board, operating under a contract the city had signed and filed and forgotten—decided which data about which people in which neighborhoods was shared with which law enforcement agencies in what form. Not a judge. Not an oversight board. Not a city council vote. Not the Rubio administration's domestic surveillance review panel, which had been convened after the Iran War and which had no authority over municipal contracts and whose 400-page report, published in 2029, had recommended "voluntary industry standards for data-sharing transparency" and had changed nothing. A company. A private company with a $4.7 billion valuation and a board that included a former CIA deputy director and a former Pentagon undersecretary and Catherine Hale, whose LinkedIn profile described her as "passionate about building safer communities through intelligent infrastructure" and whose net worth, per SEC filings, had increased by $47 million since the Austin contract was signed.

The footnote was seven lines. It negated the thirty-one pages above it. The anonymization protections, the privacy safeguards, the language the city council quoted when constituents asked questions—all of it rested on a foundation that contained an exception large enough to drive the entire surveillance system through. The wall was real. The door in the wall was also real. And NexGrid held the only key.

She went to the whiteboard. The apartment was silent—2:47 a.m., the October night pressing against the windows, the distant sound of a train on the Union Pacific tracks that cut through South Austin, the horn like a voice calling from a previous century. She wrote APPENDIX F in black marker and drew a line from NEXGRID. She wrote SOLE DISCRETION on an orange Post-it and pressed it to the board and circled it twice with the marker, the ink squeaking on the glossy surface, the sound loud in the quiet apartment. The Post-it joined thirty-one others—the investigation's nervous system, each note a synapse connecting to the next through red and blue thread that Marco had strung with the careful fingers of a man who built things by hand and who had, without understanding the full shape of what he was building, helped Elena construct a map of the most comprehensive surveillance system in American history on a $12 whiteboard from Target.

She pulled up her sensor deployment spreadsheet—the map she'd built from procurement records, infrastructure permits, city council minutes, and the public data feeds NexGrid maintained on its municipal transparency portal (which was, she now understood, transparent the way a one-way mirror is transparent: the public could see what the company chose to show them, and what the company chose to show them was not what the company was doing).

Elena had started checking her rearview mirror. Not for traffic — for Waymos. The habit had developed over three weeks, the progression so gradual she hadn't noticed it becoming compulsive until Marco pointed it out: "You checked the mirror four times between Oltorf and the river." She'd laughed. She'd said "bad habit."

She hadn't told him the truth, which was that she checked the mirror because the Waymos were everywhere and the Waymos had cameras and the cameras fed NexGrid and she was investigating NexGrid and the investigating meant that every white Jaguar on the road was a potential eye aimed at her and the potential was enough to make her pulse spike at 72 bpm every time one passed. She'd started varying her routes to The Circuit's office — three different paths, rotated randomly, the tradecraft she'd read about in journalism-security guides but that she'd never practiced until the story got close enough to feel dangerous.

The French press had become a metronome. She was drinking five cups a day — up from two, the increase correlating with the investigation's intensity, the caffeine less a stimulant than a ritual, the grinding and the pressing and the pouring a sequence her hands could perform while her mind processed footnotes. Her hands needed to be doing something. When they weren't holding a mug or a pen or a phone, they moved — tapping the desk, drumming her thigh, the restless energy of a nervous system running on cortisol and caffeine and the sustained adrenaline of a woman who'd found something enormous and couldn't tell anyone except a three-person newsroom and a man she loved and a source whose life she held in her hands. She'd lost four pounds. The thrifted blazer hung differently. The boots felt looser. She ate when she remembered to eat, which was when the hunger became a headache and the headache became a distraction from the footnotes. Marco brought her Thai food on Wednesdays. On the other six days, she ate delivery tacos and Topo Chico and whatever the French press offered, which was not food but felt like sustenance because the ritual was the closest thing to self-care she practiced.

She wasn't sleeping. Not the dramatic insomnia of a person tossing in bed — the functional insomnia of a person who worked until 2 a.m. and set an alarm for 6 and operated in the four-hour gap with the brittle efficiency of a machine running on fumes. Her eyes burned. The cheap black glasses sat on a face that was thinner than it had been in September. The grandmother's photograph inside the laptop case looked back at her with the steady, patient gaze of a woman who had survived worse and hadn't needed caffeine to do it.

She overlaid the sensor map against census data she'd downloaded from the American Community Survey. The pattern she'd suspected was now visible and undeniable—the kind of pattern that, once seen, could never be unseen, the way you can't unsee a face in a pattern of wallpaper.

Sensor density was highest in East Austin, Montopolis, Rundberg, Dove Springs—the neighborhoods where median household income was lowest and the percentage of Black and Latino residents was highest. The numbers were precise and damning: 847 sensors per square mile in the eastern corridor, versus 312 in Westlake and 401 in the Domain. The Waymo fleet's patrol density matched: the autonomous vehicles made three times as many passes per hour through East Austin as through Westlake, four times as many as through the Hill Country subdivisions west of Loop 360. The WiFi mesh was thickest where the income was thinnest. The system's attention was distributed along the same lines that every other institutional resource in Austin was distributed—inversely proportional to wealth and directly proportional to the likelihood that the people being watched lacked the political infrastructure to object.

This was the system that had scored Deshawn Watkins at 87.4%. Not because the system was racist—the code didn't contain racial bias in any line Elena could point to. The bias was in the deployment. The bias was in the density. The bias was in the decision to put 847 sensors per square mile in East Austin and 312 in Westlake—a decision made by engineers optimizing for "data richness" and approved by a procurement officer optimizing for "public safety coverage" and funded by a city council optimizing for federal grants that rewarded "comprehensive urban intelligence deployment." Each decision was rational. Each decision was defensible. The cumulative effect was a system that watched Black and Latino neighborhoods with two and a half times the intensity it watched white neighborhoods, and the disparity was invisible to every person in the decision chain because each person saw only their piece of the architecture and the architecture was designed so that no one person could see the whole.

Elena wrote RACIAL GEOGRAPHY on a green Post-it. She connected it to the sensor map with blue thread. She added WAYMO FLEET DENSITY and connected it with red.

Her personal phone — the one she used for nothing work-related, the phone she kept in a separate drawer, the phone whose browser history contained recipes and weather and the occasional Zillow listing she browsed when she fantasized about affording a one-bedroom that wasn't above a bar — buzzed on the kitchen counter. She picked it up. An Instagram ad: *NexGrid Technologies — Building Safer Cities Through Intelligent Infrastructure. Learn how our urban intelligence platform is making Austin the smartest city in America.* Sponsored. Targeted. The ad showed a smiling family walking through the Domain — glasses on, coffee in hand, the lifestyle advertisement that NexGrid used in its investor presentations. The ad was on her personal phone. The phone she never used for research. The phone that had never visited NexGrid's website, never searched for NexGrid's name, never accessed a single document related to the investigation.

She looked at the ad. She looked at the whiteboard. She looked at the three laptops on the desk where she'd spent four hours reading NexGrid's contract.

The personal phone shared a WiFi network with the three laptops. The WiFi router's DNS logs recorded every domain accessed by every device on the network. The DNS logs were collected by the ISP — Spectrum, the only provider available at her address. Spectrum's data-sharing agreements, disclosed in a privacy policy Elena had never read, permitted the company to share "aggregated browsing analytics" with advertising partners. The advertising partners included Meta. Meta's ad-targeting algorithm had detected that a device on Elena's WiFi network was accessing NexGrid-related content with unusual frequency and duration. The algorithm had inferred that the user of the network — identified by the personal phone's device ID — was interested in NexGrid. The algorithm had served the ad. The ad was not surveillance. The ad was commerce. The distinction was the lie the entire industry was built on.

She swiped the ad away. It would return. The algorithm was patient.

It was 1:30 a.m. She made coffee—the French press, the only non-electric method in the apartment, the one thing she made that didn't require a login or generate a data point. She went back to the contract.

By 3:00 a.m. she'd identified seventeen sub-sections that expanded NexGrid's authority beyond what the executive summary described, four footnotes that qualified the anonymization requirements, and a clause in Appendix G—"Indemnification and Liability Limitations"—that protected NexGrid against civil liability for "outcomes arising from the use or interpretation of platform outputs by authorized partners." If someone died because law enforcement acted on NexGrid's data, NexGrid couldn't be sued. The company had written its own immunity into the contract. The city had signed it.

At 3:15, she called James Chen.

He answered on the fourth ring, voice thick with sleep. Forty-one, former Washington Post metro reporter, started The Circuit because he believed investigative journalism could survive outside institutional media if the work was good enough and the overhead was low enough. The Circuit operated from a coworking space in East Austin with three staff: James,

Elena, and Priti the fact-checker who also managed the website. Their annual budget was less than NexGrid's monthly office snack expenditure.

"James. Sorry to wake you."

"It's three in the morning, Elena."

"I know. I found something."

"You always find something at three in the morning." But he was sitting up—she could hear the sheets, the sound of a man transitioning from sleep to the alertness that editors develop when they hear the frequency of a story that matters.

She told him. Six minutes without stopping. The contract. The appendix. The sole discretion clause. The open-ended partner list. The indemnification. The racial geography of sensor deployment. The Waymo fleet integration feeding NexGrid real-time street imagery from 4,200 autonomous vehicles the city hadn't asked for and couldn't control. James listened. She could hear him breathing.

"Who sent you to Appendix F?"

"Anonymous source. Signal. Disappearing message. I don't know who they are."

"Inside NexGrid?"

"The language in the tip was technical. 'Data-sharing addendum,' 'Appendix F.' That's platform architecture vocabulary. This is someone who knows the system from the inside."

"What do they want?"

"They didn't say. One message. One contract number. No follow-up."

James was quiet for a moment. Then: "This is the biggest thing we've ever touched."

"I know."

"If we publish and we're wrong—"

"The contract is public record. It says what it says."

"If we publish and we're right, NexGrid's legal team will try to dismantle us. They'll send a letter that costs more than our annual budget to produce."

"I know."

Another silence. "Start building. Full documentation. Everything sourced, everything verified, everything triple-checked. Don't contact NexGrid yet. Don't contact the city. Build the case first. I want to see the whole picture before we make a single call."

"Okay."

"And Elena—be careful with the source. If they're inside NexGrid, they're taking a risk we can't quantify. Protect them like they're the only source we'll ever have, because they might be."

"I will."

She hung up. She stood at the whiteboard. The red thread now connected DESHAWN WATKINS to EAST AUSTIN SENSOR CLUSTER to WAYMO FLEET INTEGRATION to NEXGRID to MUNICIPAL CONTRACT to APPENDIX F to SOLE DISCRETION. The map was growing—not the shape she'd expected, not a story about one bad algorithm or one tragic raid, but a story about an entire city that had outsourced its surveillance to a private company and given that company the unilateral authority to decide who was watched and how closely and what happened to what was seen.

She texted Marco: *Goodnight. Sorry for the late night. Love you.*

The text was sent through AT&T, which logged the metadata. Her phone sat on the desk next to three laptops and a whiteboard full of evidence. Her glasses sat beside them, cameras pointed at the ceiling, still alive, still listening in standby, the betavoltaic cell converting atoms into electricity with the patience of a device that would outlast the story, the journalist, and the system she was trying to expose.

Outside, the streetlights adjusted. A Waymo passed her building on South First at 3:47 a.m., its cameras capturing the apartment complex in the dark—the lit window on the second floor where a woman sat at a desk building a case against a system that had been designed, from its first line of code to its last line of legal language, to be unchallengeable.

Elena didn't sleep. She sat at the desk and read the contract again from the beginning, and by the time the sun came through the window at 6:40 a.m., she had forty-seven pages of notes and the beginning of something that felt like the most important story she would ever tell and the most dangerous thing she would ever do.

The anonymous message had fourteen hours left before it disappeared. Then thirteen. Then twelve. The words dissolving toward silence, the way all words should dissolve and almost never do anymore.

Elena saved her notes. She locked her laptops. She put on her glasses and went to make more coffee, and the glasses recorded the kitchen and the French press and the hands of a woman who was, at this moment, the most dangerous person in Austin—not because she had power, but because she had a footnote.

-e

* * *

CHAPTER 12

Daniel started reading documentation he had no reason to read, and the reading became a kind of addiction—the same compulsive, escalating, pattern-forming behavior that the models he'd helped build were designed to detect in other people.

It began on a Monday night, in his apartment, on the couch with the single depression on the left cushion, the MacBook Pro M10 open on his lap, a Hite beer sweating on the coffee table, the October night pressing against the windows with the particular weight of a Texas autumn that couldn't decide whether to be warm or cool and settled for both. The Samsung refrigerator hummed in the kitchen. The LG television was off but listening — its standby screen cycling through a news ticker he'd never disabled: ARTEMIS VII CREW ROTATION COMPLETES — 14 ENGINEERS ARRIVE AT SHACKLETON BASE CAMP — RUBIO CALLS FOR EXPANDED PROXY CONTAINMENT COALITION — OPENAI STOCK CLOSES AT RECORD HIGH — the information scrolling across the black screen like a heartbeat monitor for a civilization Daniel was beginning to suspect was sick. The moon had become a commute. SpaceX Starship HLS made the run quarterly now — astronauts, defense contractors, telecom engineers building the relay infrastructure, and the small crews who maintained the private facilities that had piggybacked on NASA's footprint in Shackleton Crater. Nobody watched the launches anymore. They were cargo flights. The moon was a job site. Outside, the grackles had settled into the live oaks for the night, their occasional calls cutting through the building's ambient hum like static from a radio tuned between stations.

As a platform engineer, he had legitimate access to NexGrid's technical documentation—the specifications that described how the system was designed, the data flows, the processing layers, the integration protocols that connected twelve thousand sensors to the analytical engine that processed their output. He'd read portions during onboarding. He'd written portions himself. But he'd never read the full corpus, because the architecture of a large engineering organization was designed to ensure that each person saw only the component they maintained. You optimized your module. You tested your integration. You didn't look at the whole machine, because looking at the whole machine was someone else's job—management, or legal, or a division called "Compliance and Governance" that occupied a floor Daniel had never visited and that, he was beginning to suspect, existed less to ensure compliance than to ensure that compliance was someone else's problem.

He visited it now. Not physically—digitally. He opened documents he'd never opened: integration protocols with the Austin Police Department's Real-Time Crime Center. Data-sharing agreements with federal partners. The interagency memoranda of understanding that governed how civilian data collected from streetlights and WiFi mesh nodes and Waymo autonomous vehicles became intelligence data available to law enforcement. He read the documents on his personal laptop, on his personal WiFi, wearing his Apple AI Glasses out of habit—not thinking about the fact that the glasses were recording his screen and uploading the footage to iCloud in continuous three-second buffers, which meant that Apple's servers now contained video of a NexGrid engineer reading classified municipal infrastructure documents at 1 a.m. on his couch while a Korean variety show played on the second monitor and a Zojirushi rice cooker sat on the counter and a man's life narrowed, document by document, toward a decision he hadn't made yet.

Monday night: integration protocols. The pipeline connecting NexGrid's sensor network to APD was more direct than the documentation Daniel had written for the city council suggested. The data didn't pass through a municipal review layer. It didn't wait for human authorization. It flowed automatically—alert generated by the behavioral model, transmitted to the Real-Time Crime Center, displayed on the APD dashboard, acted on by officers who saw a confidence score and a location and a recommendation and didn't know or ask how the score had been calculated, the way a patient doesn't ask how the thermometer works when the doctor says they have a fever.

Tuesday night: the Waymo integration. Daniel had known the autonomous fleet's data fed NexGrid's traffic optimization layer. What he hadn't understood was the scope. The 4,200 Waymo vehicles in Austin's metro captured twelve camera feeds each—front, rear, sides, roof—plus three LiDAR arrays generating 3D point clouds of every street, every sidewalk, every face. The data was classified as "navigation and safety mapping" and was shared with NexGrid under a commercial agreement that permitted the use of "environmental imagery for urban intelligence purposes." The Waymos photographed every person on every street they drove, and NexGrid processed those photographs through its facial recognition engine—a database that held, according to an internal metrics document Daniel found in a shared drive nobody had audited in eighteen months, over sixty million facial captures from the Austin metro alone. The city's population was 2.4 million. Every face, photographed twenty-five times. The Waymos weren't cars. They were the most comprehensive facial surveillance system ever deployed in a democratic city, and they'd been introduced as a taxi service.

Wednesday night: the body sensing. Ambient Identity Resolution—a technical specification document Daniel found buried three folders deep in a directory labeled "R&D — Pilot Programs — Municipal." The system used the WiFi mesh that blanketed Austin to identify people without cameras, without devices, without any visual input at all. WiFi signals bounced off human bodies. The reflections carried signatures unique to each person—gait pattern, body dimensions, shoulder width, stride length, the biomechanical fingerprint of a specific human being moving through radio space. The system could identify Daniel walking through the Domain without his phone, without his glasses, without his watch. Just his body in a WiFi field. The accuracy rate, per the specification: 94 percent for previously cataloged body signatures, 78 percent for new signatures matched against partial profiles.

Daniel read this and sat very still for a long time. The variety show played to no one. The refrigerator hummed. The system didn't need your phone. It didn't need your face. It needed you to exist in physical space within range of a WiFi access point—which, in Austin in 2030, meant everywhere. Every coffee shop, every office, every home, every street with a smart streetlight. You could leave your phone at home. You could take off your glasses. You could remove your watch and your AirPods and your shoes and walk barefoot and naked through the city and the WiFi would still know you by the shape of your body and the rhythm of your walk. The surveillance wasn't in the device. It was in the air. The air was the device.

Thursday night: he searched outside NexGrid. Personal laptop, VPN routed through a server in Zurich, a browser that stored no history. He searched for anyone asking the right questions—not "privacy concerns" (too broad, too vague, the language of opinion pieces that changed nothing) but "data pipeline" and "sensor integration" and "behavioral modeling" and "municipal surveillance architecture" (the language of people who understood how the machine worked, not just that it existed).

He found The Circuit on the third night of searching. A small investigative outlet he'd never heard of. Elena Ríos—four articles about Austin's surveillance infrastructure that were technically precise and journalistically careful and had been read, based on the site's public analytics, by approximately nine hundred people, none of whom appeared to have the institutional power to act on the information. She understood the landscape. She used the vocabulary. She'd filed FOIA requests that targeted the right documents. She was close to the story but didn't have the inside, and without the inside, the story was a map drawn from the outside of a building—accurate in outline but missing the rooms.

Daniel could give her the rooms.

The decision took the rest of Thursday and all of Friday and most of Saturday and the early hours of Sunday. He carried it through the standup meetings where he said "on track, no blockers" and the Slack messages he answered with the appropriate emoji and the evening jjigae that tasted wrong again—the doenjang from H Mart on North Lamar wasn't the same as the doenjang his mother bought from the vendor in the Jagalchi Market in Busan, and the difference was a fermentation culture that had been developing for forty years in a ceramic pot on a rooftop in a city Daniel couldn't visit because visiting would create a gap in the blue dot his mother checked three times a day and the gap would generate the question and the question would reveal the answer and the answer was that her son's life in America was narrowing to a point. He carried it through the variety shows that played to an apartment where the Samsung refrigerator suggested recipes and the LG television scrolled news tickers and the WiFi sensed his body on the couch and the system classified his evening as "low activity, residential, consistent with baseline" — the system's language for a man sitting very still because the stillness was all that kept the decision from arriving before he was ready to receive it.

He carried it to the Korean Presbyterian church on Sunday morning—a small congregation in a converted office building off Metric Boulevard, thirty-seven people, the hymns sung in Korean by voices that didn't need a key to find the pitch because the songs were older than the building and the building was older than the algorithm. The blue dot on his mother's Samsung Galaxy Glass 3 showed him present—the dot she checked at 9 a.m., at 1 p.m., and at 6 p.m. Busan time, the three-check pattern that was her surveillance system, her only surveillance system, a mother's daily verification that her son was alive and in Austin and the dot was in the Domain and everything was fine. The dot didn't know Daniel was in a church pew wondering whether to destroy his career. The dot didn't know Daniel had a screenshot on his phone that could dismantle a municipal surveillance infrastructure. The dot knew his coordinates. It did not know his heart.

The pastor preached about obedience. Following God's law even when it conflicted with the law of men. Daniel sat in the back row and listened and thought about routing logs and confidence scores and a twenty-two-year-old man who'd coached basketball and whose notebook full of plays was in an evidence bag that nobody had opened. He thought about the 87.4%. He thought about the 4.2 percentage points from a phone call to a grandmother. He thought about his own phone calls to Busan—Tuesday and Thursday at 6:30 p.m. Austin time, 8:30 a.m. Busan time, the pattern so consistent it was practically a beacon. If someone ever flagged his communication metadata, the calls to South Korea would contribute to a score the same way Deshawn's calls to Lagos had contributed to his. The architecture was identical. The geography was different. The outcome, if the system ever decided Daniel was a threat, would be the same.

After church, he walked through the Domain. Past the Sweetgreen playing Brooks Rosser from the outdoor speakers. Past the Whole Foods with its "Just Walk Out" cameras. Past the Tesla showroom and the Lululemon and the Warby Parker where a 3D scan of his face sat in a database. The grackles were screaming from the live oaks—the evening chorus, hundreds of birds producing a cacophony that sounded like a server farm crashing, the one sound in the Domain no designer had curated and no algorithm could suppress. He walked past all of it without looking up.

He carried it home. He heated the jjigae. He hummed the ballad. He sat on the couch.

The decision came the way decisions come when they've been building for days—not as a sudden clarity but as the exhaustion of alternatives. He couldn't unknow what he knew. He couldn't keep carrying it alone. He couldn't go to his team lead or to Catherine Hale or to the Compliance division because every path inside NexGrid led to lawyers who would protect the company and to HR personnel who would protect the lawyers and to a security team that would protect the system, and the system was the thing that had killed Deshawn Watkins, and Daniel was part of the system, and the only way to stop being part of the system was to become a threat to it.

He opened Signal. He'd downloaded it weeks ago, configured it the way NexGrid's own security training taught: VPN active, disappearing messages enabled, no profile photo, no linked number. The company had trained him to be invisible. He was using the training against the trainer.

He typed the message. One sentence. A contract number and a direction: Appendix F. He'd chosen the entry point with an engineer's precision—a publicly available document that required no classified material to access, that would give Elena the legal framework without exposing Daniel as the source. Once she read it—the sub-clauses, the footnotes, the sole discretion clause in eight-point font—she would see the architecture of authority the city had given NexGrid. She would start asking what the system was actually doing with that authority. And that question would lead to the data pipeline and the bypassed anonymization and the confidence scoring model and the eastern corridor and Deshawn Watkins and everything else that Daniel couldn't carry by himself anymore.

He stared at the message. The cursor blinked. The apartment was silent—no variety show, no game, no music. Just the refrigerator and the HVAC and the beating of a heart that his Apple Watch Ultra 5 was recording as 89 bpm, flagged as "elevated resting, recommend wellness check," an anomaly the system would note and file and never understand.

He pressed send.

The message left his device, traveled through encrypted channels to a server in Switzerland and from the server to Elena's phone in Austin, covering the distance between two apartments in the same city in 1.3 seconds. The message said almost nothing. It said enough.

He put the phone down. He sat in the dark. He didn't feel brave. He felt the specific lightness of a man who has set down something heavy and hasn't yet calculated what the setting down will cost. The cost was coming. He knew it the way he knew routing architecture—structurally, in the logic of systems. His access patterns would be reviewed. His reading history flagged. The question would be asked: "Why were you looking at these documents?" And the answer he gave would determine whether Daniel Yun continued to exist in the country that was the only place his parents could reach him by phone and track him by dot and believe, from seven thousand miles away, that their son was safe.

보고 싶다, he thought. I want to see you.

He didn't know if he was thinking of his mother or the truth or the country he'd left or the version of himself that existed before the screenshot. All of them. All of them gone or going.

The message on Elena's phone was counting down. Twenty-four hours. The words dissolving toward silence, toward the erasure the system promised and the infrastructure denied. The words would disappear. The data—the routing logs, the confidence score, the 87.4 percent, the system performance notes that described a man's death as an output within acceptable parameters—would not.

Outside, a Waymo passed the building. Its twelve cameras captured the facade, the windows, the streetlight adjusting for nothing. The man on the fourth floor who had just sent a message that could dismantle a piece of the system was invisible behind curtains the cameras couldn't penetrate.

Not yet.

-e

***

CHAPTER 13

The colleague's comment came on a Tuesday, in the break room, over a microwave burrito that smelled like every microwave burrito in every government break room Renée had ever worked in—the universal scent of institutional lunch, of budgets spent on servers instead of people, of Styrofoam cups of weak coffee drunk under fluorescent lights by professionals who couldn't discuss what they did with anyone outside the building.

"Hey, Renée—have you noticed the system running access audits more frequently? I got flagged for a search I did last month on that Bexar County integration. Routine, they said, but still. Had to fill out a justification form."

Linda. Two desks over. Intelligence support analyst. Competent, reliable, bad at reading rooms, and apparently—in this one instance—accidentally devastating in her timing, the way a person who doesn't know about the crack in the dam can cause a flood by leaning against the wall.

Renée's thumb found her opposite palm beneath the break room table. She pressed. The pressure was grounding—a physical anchor against the wave of adrenaline she could feel building in her chest, the specific cold rush of a compartment under stress. "Yeah, they've been tightening up across the board. Post-incident protocol. It'll cycle through."

"Makes sense. Just annoying. Like, I pulled a public county record and I have to justify it? Come on." Linda unwrapped the burrito. The microwave-warmed tortilla split and beans leaked onto the Styrofoam plate. "Anyway. Did you see the new SIGINT brief?"

"Not yet."

"Three new names. Southwest corridor. All financial."

"I'll look at it."

The conversation lasted forty seconds. Renée carried it for three days.

She didn't know if the audits were about her. She might never know—that was the particular cruelty of surveillance within a surveillance organization: the watchers were watched by systems they couldn't see, using methods they'd helped design, and the uncertainty was itself a form of control. The facility ran access reviews on rotating schedules—quarterly for routine systems, monthly for classified platforms, continuous for the highest-sensitivity databases where Renée had been running queries she had no operational reason to run. The reviews were automated. A pattern-detection algorithm—structurally similar to the one that had scored Deshawn Watkins—scanned access logs for behavioral anomalies: unusual search terms, atypical access times, queries that deviated from an analyst's established profile. The system watched the watchers. That was the design. That was always the design.

Renée's access patterns over the past three weeks were, by her own professional assessment—and she was a very good professional—flaggable. Seventeen queries on systems she wasn't actively assigned to. The Deshawn Watkins provenance document, pulled from an interagency file she had authorization to access but no operational reason to open. The behavioral model methodology—weights, thresholds, input categories—available at her clearance level but not something an intelligence analyst reviewed unless working on a specific assessment. And the queries for Amara's phone number, her device ID, and the address of her apartment—three searches that individually could be explained as database familiarization and collectively painted the portrait of a woman using classified tools for unauthorized purposes. The portrait of a federal employee committing the specific crime of unauthorized computer access, 18 U.S.C. § 1030, which carried a maximum sentence of ten years and would end not just her career but her freedom and her ability to protect the person the crimes were committed to protect.

She'd been careful. She'd spaced the queries across sessions. She'd interspersed them with legitimate work—the SIGINT briefs, the southwest corridor assessments, the financial monitoring that constituted her actual job. She'd never queried Amara's name directly, only the phone number and device ID, because a name search left a trace that was harder to explain than a number search, which could always be attributed to investigating an unknown contact. She'd used the tradecraft she'd learned over twenty years—the techniques of a woman who understood surveillance intimately enough to know its blind spots.

But the blind spots were shrinking. The monitoring systems were newer, more granular, more capable of detecting exactly the kind of subtle behavioral shift her queries represented. Three years ago, her pattern might have been invisible in the noise of thirty-two analysts running thousands of queries daily. Today, with post-incident audit protocols elevated in the wake of a tactical operation that had ended in a man's death, with the system's sensitivity dialed up and the review algorithms retrained on tighter parameters, the margin was thinner than it had ever been.

She went about her days. She did her work. She was professional, thorough, precise. Her reports were clean. Her briefings were accurate. She smiled at Linda in the break room and said "morning" to the section chief and participated in the Wednesday all-hands meeting with the engaged attentiveness of a woman who had nothing to hide and everything to lose. The performance was flawless. It had always been flawless. Renée had been performing competence and composure in rooms that would have destroyed lesser people since she was nineteen years old at Fort Huachuca, a skinny girl from Houston who'd enlisted because the Army offered a path and her parents' prayers alone couldn't fund a future.

The performance held. Inside it, the fear lived like a second heartbeat.

She ran harder that week. Seven miles, then eight. In the dark, 4:50 a.m., the old Asics on pavement, phoneless, glassless, the Waymos passing like white ghosts—their cameras capturing her silhouette, her gait, her body moving through predawn streets that belonged to no one but felt, for forty-seven minutes, like they belonged to her. She ran until the thinking stopped and the body took over and the world reduced to breath and pavement and the rhythm of her feet striking the ground in a cadence older than any algorithm, more reliable than any database, a rhythm her body had known since before she knew what intelligence meant or what it could do to the people it was pointed at.

She prayed while she ran. Not in words—in motion. The running was the prayer. The expenditure of the body was the offering. She didn't form sentences addressed to God. She moved, and the moving was a conversation with something she couldn't name: I am carrying more than I can carry. I am holding more than I can hold. I need you to take the part I can't hold, because I can't put it down, and if I drop it, the girl is lost.

***

Sunday morning. Grace Fellowship Baptist Church. MLK Boulevard.

The October sun hit the stained glass at 9:30 and threw colored light across the pews—red and blue and gold patterns that moved with the hour, the oldest technology in the building, light through glass, requiring no battery and no login. The sanctuary smelled of wood polish and the perfume of forty women and the faint sweetness of the communion grape juice that sat in tiny cups on a table near the altar, covered with a white cloth. Someone had brought kolaches from the Czech Stop in West—a three-hour round trip—and the box sat on the welcome table in the lobby alongside bulletins and visitor cards and a QR code for the church's giving app.

The building had been a Methodist church in the 1960s, a community center in the 1980s, and a church again since 1997, when Reverend James Okonkwo—Nigerian-born, Houston-raised, a man whose sermons lasted exactly forty-two minutes and covered the distance from Genesis to the evening news without seeming to rush—gathered a congregation from the Nigerian diaspora and opened the doors. The sanctuary held 180 people. The pews were original Methodist, dark wood, worn smooth by seventy years of Sunday weight. The sound system was modern—a QSC array controlled from an iPad at the back. The WiFi was open, no password, named "GraceFellowshipGuest," logging every device that connected. Two security cameras—installed after a break-in, the footage stored on a cloud server maintained by a company that also provided security systems to ICE detention facilities, a connection the church board had never asked about and the security company had never disclosed.

Renée sat in her usual pew. Third from back, left side. Dark dress, low heels—the church outfit, different from the work outfit, different from the running outfit, each set of clothes a uniform for a different compartment of a life built on compartments. Her Apple AI Glasses were on—she'd considered leaving them in the car. She'd decided against it. Arriving without glasses would draw exactly the kind of concern she couldn't afford: "Are you okay? Is something wrong?" The concerned voices of a congregation that noticed when someone deviated from routine. The glasses stayed on. Sixty pairs of cameras in a sanctuary, recording the service, the preacher, the grief and faith and singing. The word of God captured in sixty cloud accounts and processed by platforms that would analyze the footage for engagement metrics and targeted advertising and never once recognize it as worship.

Reverend Okonkwo preached. His text was from the book of Daniel—the prophet, not the engineer—chapter three. Shadrach, Meshach, and Abednego in the fiery furnace. Three men who refused to bow to the king's golden image. Three men who chose the fire over compliance. Three men who trusted that God would be in the furnace with them, even if the furnace consumed them.

"Even Pope Leo — our Catholic brother in Rome — said at Easter this year that the furnace of this century is not made of fire. It is made of convenience." Reverend Okonkwo paused, letting the reference land. He quoted ecumenically when the source was good. "The Holy Father said we have traded our freedom for comfort and called it progress. I say we have traded our souls for Wi-Fi and called it connection." A murmur of amens. Someone's phone buzzed in a purse — a USA Today push notification: IRAN PROXY CELL CLAIMS INFRASTRUCTURE ATTACK IN THREE EU NATIONS — and the woman silenced it without looking, the war's ongoing aftershocks absorbed and dismissed as ambient noise, the way the congregation absorbed and dismissed the Waymos passing outside the stained glass.

"The furnace is not the absence of God," Reverend Okonkwo said. His voice was deep, carrying the particular musicality of a man who'd grown up speaking Igbo and English in alternating sentences and had brought the rhythm of both into his preaching. "The furnace is where you meet God. Not in the comfort. Not in the safety. In the fire. When you have made the choice the world calls foolish. When you have refused to bow to the thing everyone else is bowing to. When you are standing in the heat and the heat is real—that is where God meets you. Not before. Not after. In the fire."

Renée listened. She'd heard this sermon before—the furnace was a staple, a text Nigerian preachers returned to because the story of resistance under imperial pressure resonated with a people whose history included colonialism and dictatorship and the daily navigation of building a life in someone else's country. She'd always heard it as metaphor. Today she heard it as operational briefing.

She was in the furnace. She'd refused to bow—not to a golden image but to a system, an algorithm, an institutional architecture that claimed to see everything and understood nothing and would, if she stopped interfering, process her niece into a database and from the database into a deportation and from the deportation into a country Amara hadn't seen since she was eleven and didn't remember and couldn't survive. Renée had chosen the furnace. She'd chosen Amara. The choice was illegal by every standard of the institution she served and entirely righteous by every standard of the faith she held, and the contradiction was the fire, and the fire was real, and she was standing in it every day she went to that workstation and ran those queries and drove those varied routes and pressed her thumb into her palm and prayed without words that the God of Shadrach and Meshach and Abednego would be in this particular furnace too.

The congregation sang. Renée sang with them—a hymn she knew by heart, her voice steady, her face composed. No one in the pew around her would have guessed that the woman singing "Great Is Thy Faithfulness" was simultaneously calculating the probability that her access logs had been flagged and estimating the number of working days before the investigation reached the queries that would reveal what she'd been doing and for whom.

The service ended. She shook hands. Said "God bless" to three people and "see you Wednesday" to two and "tell your mother I said hello" to one—the small liturgy of community that she performed every Sunday with the sincerity of a woman who meant every word and understood that the words were also cover, also performance, also the sound of a woman who had nothing to hide standing in a parking lot where two Waymos passed in three minutes, their cameras capturing the church, the congregation, the license plates, the faces.

She sat in the Camry. She did not start the engine. The parking lot emptied around her—cars pulling out, each logged by the security camera, each plate scanned by the Flock Safety reader on the utility pole at the corner of MLK and Chestnut. She watched the cars leave and thought about furnaces and golden images and the moment in the story that nobody talked about—the moment before the fire, when the three men stood in front of the king and the king gave them one last chance to bow and they looked at each other and made the decision together.

Together. That was the part that broke her.

Renée had made the decision alone. She'd been making it alone for three years—since Amara turned eighteen and the thin protections of childhood evaporated and Renée had started accessing databases she shouldn't access to check for flags that shouldn't be there. Three years of solitary resistance. Three years of furnace. No Meshach. No Abednego. Just Renée in a parking lot with a Bible in her lap and a classified workstation waiting for her on Monday and a niece cooking jollof in an apartment the system hadn't found yet.

She started the car. She drove home. A different route. Always a different route.

The house was clean and empty. She changed out of the church clothes. She made tea—the electric kettle, the only smart device she'd purchased voluntarily, and only because boiling water on the stove took too long and the kettle's WiFi connection was a compromise she could live with. She sat on the edge of her bed and opened the Bible.

Daniel 3:17-18. *If it be so, our God whom we serve is able to deliver us from the burning fiery furnace, and he will deliver us out of thine hand, O king. But if not, be it known unto thee, O king, that we will not serve thy gods, nor worship the golden image which thou hast set up.*

But if not.

The most terrifying three words in the Bible. The acknowledgment that God might not rescue you. That the furnace might consume you. That faith is not a guarantee of safety but a commitment to stand anyway, burning anyway, because the alternative—bowing to the image, complying with the system, letting the machine process the people you love into the categories it was built to process them into—is worse than the fire.

Renée closed the Bible. She picked up her phone. She opened the music app and played the Yoruba hymns —not quietly this time, not under her breath, not at the barely audible volume she usually maintained as if ashamed of the sound. She played them at full volume in her empty house, the music filling the rooms the way her mother's voice used to fill the kitchen in Houston on Sunday afternoons, the melodies carrying across the hallway and into the living room and against the windows where the afternoon light came through and turned the glass to gold. The music bypassed every compartment. It found the version of Renée that existed before the uniform, before the clearance, before the divorce, before the choices. It found the girl in Houston whose mother sang while she cooked and whose father prayed with his hands on the steering wheel in traffic and whose world was small and warm and watched by nothing but God.

She sat on the edge of her bed and pressed her thumb into her palm and let the music carry the weight she couldn't carry alone.

But if not.

She would not bow. Whatever came on Monday—the audit, the flag, the conversation with the section chief that began with "Renée, we need to talk about your access patterns"—she would not bow. She would walk into the furnace because Amara was in the furnace and God was supposed to be in the furnace too, and if He wasn't, if the fire took everything, then the fire took everything, and she would have chosen correctly even if the choice destroyed her.

The music played. The house was full of sound and empty of people. No Graham, no Alexa, no smart TV—just a woman and a Bible and hymns from Lagos in a house with the bare minimum of devices the century required and the maximum amount of faith one person could hold without breaking.

The glasses sat on the dresser. Cameras dark. Betavoltaic cell still running in the temple, converting atoms into electricity in the silence. The device would outlast the woman. The battery would outlast the faith. The system would outlast them all.

Outside, a Waymo passed on Berkman Drive. Its cameras captured the house, the car in the driveway, the dark windows, the afternoon light. It didn't know who lived here. It didn't know about the Bible or the music or the niece or the furnace. It added the photograph to the map and kept driving, because that was what it did—it captured everything, understood nothing, and kept driving.

Part Two ended where it began—with a system watching and a woman being watched and the distance between them narrowing with every day she chose not to bow.

-e

***

-e

PART THREE — "Exposure"

CHAPTER 14

The FOIA response came at 6:12 a.m. on a Tuesday in late October, fourteen weeks after Elena filed the request. She was still in bed. The phone lay on the pillow beside her face, close enough that she felt the buzz before she saw the light. On the nightstand, the Apple AI Glasses waited with their cameras dark, the betavoltaic cell humming its endless low note in the silence of a South Austin apartment that smelled like last night's cold coffee and the citrus ghost of the Topo Chico she'd killed at 2 a.m. reading court filings. Fourteen weeks of waiting. And the city had answered before she was even awake to catch it.

The email contained a single PDF attachment—4,712 kilobytes, 412 pages—and a cover letter that began with an apology for the delay ("Due to the volume and complexity of the responsive documents, additional time was required to complete the review process") and ended with a sentence that would have been funny if it wasn't criminal: "Certain responsive documents have been withheld or redacted pursuant to Section 552.101 of the Texas Government Code, which protects information considered confidential by law."

Confidential by law. Elena read this on her phone's screen, lying on her side in the gray pre-dawn light that came through the window and fell across the whiteboard on the opposite wall—the whiteboard that had grown over three months from a few Post-its and a single red thread into a map that covered two-thirds of the surface, the taxonomy of colored notes expanding like a city's sprawl: yellow for confirmed facts, pink for unconfirmed leads, green for legal questions, orange for sources. The light illuminated the board in a glow that made the investigation look, for a moment, like a stained-glass window in a church she didn't attend. She sat up. She swung her legs to the floor. She did not put on her glasses—the morning rebellion she maintained against the century's insistence that every waking moment be recorded by the camera you wore on your face.

She walked barefoot to the desk—the Ikea desk whose hydraulic chair sank two inches over the course of each workday, so that by evening Elena was sitting lower than she'd been at morning, a slow descent she'd stopped noticing the way she'd stopped noticing most of the small erosions the years had produced. She opened the MacBook. She downloaded the PDF.

She made coffee first. The French press—the only non-electric appliance in the apartment, the one object that didn't require a login or generate a data point or transmit its usage to a server that would store the information alongside the usage patterns of 14 million other customers. She ground the beans by hand with a ceramic burr grinder she'd found at a garage sale in the Valley three years ago, the kind her grandmother had kept on a shelf in the kitchen where the light came through a window that faced the desert. The rhythmic crunch of the handle was the most analog sound in Elena's life—repetitive, physical, producing a result she could smell. The kitchen filled with dark roast and October morning air through the cracked window: 58 degrees, the cool front that had pushed through overnight, the first real chill of fall, the air carrying cedar from the Hill Country and the faint mineral scent of the Colorado River and the distant hum of I-35 beginning its daily acceleration from silence to roar.

She poured the coffee into the mug with the dried espresso ring she never fully cleaned—the mug that had been on this desk through every FOIA appeal, every 2 a.m. document review, every cooking video she watched to calm down when the investigation made her hands shake. She sat down. She opened the PDF.

For three hours, Elena Ríos read the document that would change the shape of her life.

The 412 pages were the complete NexGrid municipal contract—not the public version available on the procurement portal, not the executive summary she'd already analyzed and annotated and taped to the whiteboard in six overlapping pages, but the full agreement including all appendices, amendments, addenda, and the supplementary data-sharing protocols that the public version referenced in footnotes but did not contain. The difference was the difference between reading the menu at a restaurant and reading the health inspection report in the kitchen. The public version said NexGrid provided urban intelligence services to the City of Austin. The full version said how, and to whom, and under what authority, and with what data, and the answers occupied 412 pages of language so dense and so deliberately impenetrable that Elena understood, by page 30, why it had taken fourteen weeks to process. The city's legal team had reviewed every page before releasing it. They'd redacted eleven passages—black bars across sentences she'd later try to reconstruct from context. They'd released the rest because the rest was, technically, public record, and because the language was so expertly constructed that a non-specialist reader would understand none of it and a specialist reader would need weeks to decode the cross-references and defined terms and nested exemptions that turned every paragraph into a puzzle whose solution revealed another puzzle underneath.

Elena was a specialist. She read the way she'd been trained to read government documents—slowly, with a digital highlighter, pausing at every defined term to check the glossary in Appendix A, tracking cross-references backward and forward, building a mental map of what the document said underneath what it appeared to say. The October sun moved across the floor as she read. It reached the whiteboard at 8:15 and set the red thread glowing. It reached the desk at 9:30 and warmed her coffee hand. It climbed the far wall by 10:00 and disappeared behind the bookshelf. She didn't notice any of it. A Waymo passed her building four times during the three hours—four passes, forty-eight camera feeds, the system photographing her building at regular intervals the way a guard walks a perimeter. She didn't hear the electric hum. The French press went cold. The apartment contracted to the dimensions of the screen.

Appendix F she already knew—the sole discretion clause, the open-ended list of authorized partners, the footnote in eight-point type that gutted the anonymization protections. But the full document contained Appendices H through N. Seven bilateral data-sharing agreements with specific federal agencies. Each one a separate contract within the contract—a Russian nesting doll of legal authority, each layer granting broader access than the one above it suggested.

Appendix H: U.S. Immigration and Customs Enforcement. NexGrid would provide "processed behavioral intelligence outputs including device proximity profiles, facial capture confirmations from autonomous vehicle integration feeds, and WiFi-based identity resolution data" for individuals "identified through multi-modal analysis as potentially subject to immigration enforcement actions." The language was surgical. It meant: NexGrid would use the Waymo fleet's cameras and the WiFi body-sensing system and the device proximity logs to identify undocumented people and hand the data to ICE. Elena thought of every taco truck on East Riverside, every bus stop where women with no papers waited for routes that took them to jobs that paid in cash, every apartment complex where families lived three generations deep because the lease was in one name and the bodies were many and the system that counted them had just been given permission to count.

Appendix I: FBI Joint Terrorism Task Force. Real-time alert forwarding for confidence scores exceeding designated thresholds—the same pipeline that had generated 87.4% on Deshawn Watkins, the same architecture that converted a phone call to Lagos into 4.2 percentage points of threat probability. The appendix referenced "methodology validated in theater deployment, 2026-2027"—the Iran War. The behavioral scoring model NexGrid used on the streets of Austin had been deployed in the war theater. The facial recognition that photographed pedestrians from Waymo cameras had photographed targets from military drones. The WiFi body sensing that tracked Renée through her morning run had tracked insurgents through the streets of Tehran. The technology had been developed for war and licensed for peace, and the licensing agreement was in the contract, and the contract had been signed by a city that had never been told its public safety platform was built on military surveillance architecture.

Appendix J: Drug Enforcement Administration. "Financial anomaly detection outputs" shared with the DEA's El Paso Intelligence Center—the same center that coordinated drug interdiction along the southern border, the same infrastructure Elena's grandmother had watched from the wrong side of the fence. The data flowing from Austin's streetlights to the DEA included cryptocurrency transaction patterns, device proximity correlations at flagged commercial addresses, and "behavioral indicators consistent with narcotics distribution activity"—a phrase that could mean a dealer on a corner or a teenager buying cryptocurrency to pay for something private on a browser he thought was anonymous.

Appendix K: the full Waymo integration protocol. Not traffic optimization. Surveillance. Four thousand two hundred autonomous vehicles, twelve camera feeds each, the data classified as "navigational imagery" for contractual purposes but processed through NexGrid's facial recognition engine—a database of sixty million captures from the Austin metro alone—and matched against law enforcement databases in real time. The Waymos photographed every face on every street they drove. Every face. Every street. Every hour. The cumulative capture rate meant that every person in Austin's metro area had been photographed, on average, twenty-five times. The data flowed through a pipeline governed not by a warrant but by a commercial licensing agreement, because NexGrid was a private company and private companies could buy whatever data they wanted and sell whatever data they collected. The legal architecture was elegant and terrifying: the Fourth Amendment protected citizens from government surveillance, but it said nothing about a corporation that surveilled citizens and then sold the data to the government. The founders hadn't anticipated the arrangement because the founders hadn't anticipated a world where the corporation and the government were connected by a contract and the contract was hidden in a procurement database and the database was public and nobody searched it.

Appendix L: WiFi Ambient Identity Resolution. The body-sensing system described with the clinical precision of a medical paper—gait analysis, dimensional mapping, biomechanical signature extraction, the technology that could identify a specific human being by the way their body interrupted a WiFi signal. Anonymization was described as default. A footnote—number 41, page 287, eight-point type—noted that body signatures could be "correlated with device-based identifiers when multi-modal confirmation is available." The anonymization was a door. The footnote was the key that opened it from the inside.

And then Appendix M. Three pages. Titled "Off-Site Data Archival and Jurisdictional Continuity."

Elena read it four times. Each time she believed it less and understood it more.

Appendix M described a data storage facility operated by a NexGrid subsidiary called Selene Infrastructure Partners, located on the lunar surface. Shackleton Crater. South pole of the moon. The facility stored "high-sensitivity processed intelligence outputs"—the unmasked profiles, the behavioral confidence scores, the raw surveillance data that had been stripped of anonymization under the sole discretion clause—transmitted via encrypted satellite uplink from NexGrid's terrestrial servers to a relay satellite in lunar orbit and then beamed to the surface facility. The contract characterized the jurisdictional framework as "extraterrestrial data sovereignty" and noted that data stored at the Selene facility was "maintained outside the territorial jurisdiction of any signatory to the Outer Space Treaty of 1967 and is not subject to domestic legal process including but not limited to subpoenas, search warrants, and Freedom of Information Act requests."

The most sensitive surveillance data NexGrid collected about the citizens of Austin, Texas—the data that killed Deshawn Watkins, the data that mapped Amara's walks to the bus stop, the data that tracked Daniel's money transfers and Craig's hotel visits and Elena's own FOIA research—was being stored on the moon. 238,900 miles from the nearest courthouse. Beyond the jurisdiction of any court on Earth. Beyond the reach of any subpoena Elena could file, any congressional committee could issue, any judge could enforce. Beyond the physical ability of any human being to access without a rocket and a spacesuit and the cooperation of the company that had built its memory in the one place where no law could demand its return.

Elena sat back in the sinking chair. Her hands were cold—not from the temperature but from the adrenaline, the blood retreating from her extremities the way it retreated during a car accident or a fall, the body's ancient response to a threat the mind hadn't finished processing. Outside, a grackle screamed from the power line—the mechanical shriek cutting through the apartment's silence like an alarm that nobody else could hear. A Waymo glided past on South First, its twelve cameras capturing her building, her window, the glow of her laptop screen visible through the glass. The system that stored its data on the moon was photographing her apartment from the street.

She called James Chen.

He answered on the first ring. He'd been tracking the FOIA timeline the way he tracked every timeline—with the anxious precision of an editor whose newsroom employed three people and whose annual budget was less than NexGrid's monthly office coffee expenditure.

"You got it," he said. Not a question.

"I got all of it." She told him. Twenty minutes. The Waymo integration. The WiFi body sensing. The federal appendices—ICE, FBI, DEA, and two agencies designated by four-letter codes she couldn't match to any publicly known entity. And Selene. The moon.

James was quiet for a long time. She could hear his kitchen—the whir of a coffee grinder, morning sounds, the ambient architecture of a man's apartment absorbing a piece of information that changed the shape of the room it occupied.

"The moon," he said.

"The moon."

"How do we verify the facility exists?"

"SEC filings for Selene Infrastructure Partners. Corporate registration—they'll be in Delaware, that's where you incorporate aerospace subsidiaries. FAA commercial launch manifests. If NexGrid launched a payload through SpaceX, there's a record. And James — the facility is at Shackleton Crater. Three kilometers from the Artemis Base Camp. There are fourteen engineers up there right now on the current rotation, building habitat modules and maintaining relay infrastructure. NexGrid's data center is sitting next to the place where Americans go to work every quarter."

"You're telling me there are people commuting to the moon — people who applied on LinkedIn, people whose kids FaceTime them through the Gateway relay — and they're working three klicks from a private server that holds the behavioral profile of every person in Austin."

"That's what the contract says."

"Start pulling it. Don't contact NexGrid. Don't contact the city. Don't tell anyone outside this call." He paused. "Including Marco."

"I know."

"Elena. Your source. The Signal message. They sent you to the document that leads here. They knew about Selene."

She thought about the anonymous message—the sixteen-character endpoint Daniel had found in the routing log, the satellite uplink, the data flowing from East Austin to a destination that didn't appear in any documentation. The source had found the pipeline. The source had traced the data to the edge of the earth and couldn't follow it further. So the source had sent Elena the contract that explained where the edge led.

"I think my source found the uplink," she said. "I think they found data flowing off-world and couldn't carry the weight of it alone. So they sent me the map."

"And the map leads to the moon."

"The map leads to the moon."

She hung up. She went to the whiteboard. She took the red marker—the same marker she'd used for three months, the tip slightly dried, the ink the color of the thread that held the investigation together—and wrote SELENE on a new Post-it. She pressed it to the board. She drew the line: from NEXGRID through APPENDIX M to SELENE to a new note she labeled SHACKLETON CRATER. She circled it twice, the marker squeaking on the glossy surface.

The whiteboard now contained a map that began with a dead man on a street in East Austin and ended on the moon. The red thread connected them through twelve intermediary nodes—sensor clusters, data pipelines, confidence scores, federal agencies, corporate subsidiaries, satellite uplinks, a commercial launch vehicle that had carried a 2,400-kilogram payload to the south pole of the moon fourteen months ago—and the map described a system that had been designed, from its first line of code to its last line of legal language, to collect everything about everyone and store the most sensitive data in a place where no law could touch it.

She stepped back. She looked at the map. She looked at the line that ended at the moon. She thought about the night sky over the Valley—the sky her grandmother had looked at from a porch with a chain-link fence and a dog that barked at the Border Patrol trucks and a garden that grew tomatoes and peppers and the kind of stubborn hope that produces vegetables in desert soil. The same moon. The same light. The same rock in the sky that humans had looked at for 200,000 years and seen mystery and longing and God and that now held the browsing history of a dead man and the confidence score that killed him and the floor plan of every house whose owner had bought a vacuum cleaner that mapped the rooms while it cleaned them.

She opened her laptop case. She looked at the grandmother's photo—faded, fragile, taped to the inside of the lid. A woman standing in the Texas sun with her hand on a chain-link fence, squinting into light the camera had captured but the photo couldn't convey. The one image in Elena's life that existed in a single copy. Not backed up. Not in any cloud. Not on any server or any moon. An artifact so fragile that a glass of water could destroy it, and for that reason, the most precious thing she owned.

Everything else was on the moon now. And the moon wasn't giving it back.

She closed the case. She put on her glasses. She opened the SEC database on one laptop, the Texas Secretary of State's corporate records on another, and the FAA's commercial space launch database on the third.

Outside, October in Austin. The grackles screamed from the power lines—the one sound nobody designed, nobody curated, nobody could suppress. The Waymos glided past on South First, photographing everything, understanding nothing. And a thirty-one-year-old woman in a sinking chair at a $79 desk with a cold French press and a whiteboard and a photograph of her grandmother was about to prove that a private company had put the private lives of 2.4 million people on the moon, one footnote at a time.

-e

* * *

CHAPTER 15

The Bluetooth speaker in the Holloway kitchen was a Sonos Era 100, mounted beneath the cabinet beside the microwave, connected to every device in the house through the Samsung SmartThings mesh that coordinated the fifty-one networked objects the family had accumulated over three years of purchasing things that promised to make life easier and that had, instead, made life observed. The speaker had been configured to auto-pair with the nearest phone when someone entered the kitchen—a feature Megan had enabled because she liked her podcast to follow her from room to room, the episodes of true crime narration traveling with her body through the house the way the WiFi tracked her body through the signal. The feature did not discriminate between phones. It connected to whichever device arrived first, because Bluetooth was democratic in the way that all surveillance technologies were democratic: it treated everyone the same way, which was to say it treated everyone as a signal source.

On a Thursday evening in the third week of October, Craig walked through the garage door into the kitchen with his phone in his pocket and his AirPods Pro 6 in his ears and his mind on Rachel—specifically on the text she'd sent at 4:47 p.m., a text he'd read at a stoplight and deleted at the next one, a text that said *thinking about Saturday* in the language of a woman who didn't need to say more—and his guard was down for the first time in eleven months. The Sonos detected his phone's Bluetooth signature before his AirPods could claim priority. The audio handoff happened in 0.3 seconds. For four seconds, the Holloway kitchen filled with Tupac Shakur—"Dear Mama," the opening bars, the piano, the bass deep enough to vibrate the water glasses on the shelf beside the refrigerator—before Craig's hand found his phone in his pocket and killed the connection.

Four seconds.

It was enough.

Megan looked up from the island where she was reviewing Etsy orders on the refrigerator's 32-inch touchscreen, her rose-gold Warby Parker glasses reflecting the screen's glow. Tyler, at the counter eating a bowl of Honey Nut Cheerios with the mechanical efficiency of someone who wanted to be elsewhere in ninety seconds, looked up for the first time in what Megan estimated was three days—not because the music interested him but because the sound of Tupac coming from the family kitchen speaker was so incongruous with the Holloway household's curated acoustic identity that his nervous system registered it as an anomaly before his conscious mind did. Brooke, passing through the kitchen with a glass of water, stopped in the doorway. Her tortoiseshell glasses were pushed up on her forehead—the off-the-record gesture—and her hazel eyes were doing what Brooke's eyes always did: watching.

"What was that?" Megan said.

"Podcast. Sorry. Bluetooth grabbed the wrong speaker." Craig's voice was the voice he used for explanations—easy, fluent, the same register he deployed at showings when a client asked about a water stain or a settling crack. He'd been lying about small things for so long that the lies had the texture of truth, the way a path worn by daily use stops looking like a detour and starts looking like the road.

Megan looked at him for one second. One second that contained the full weight of a woman who had heard eleven months of small lies and was beginning to hear them not as individual deceptions but as a frequency—a constant hum beneath the marriage, like the sub-audible tone of the HVAC system that you stopped noticing until someone turned it off and the silence revealed how loud it had been. She turned back to the screen.

Tyler went back to his cereal. He didn't look up again. His AirPods were in—noise cancellation active, the world sealed out. The four seconds of Tupac had been absorbed and dismissed.

Brooke looked at Craig. She held his gaze for two seconds—two seconds during which the Holloway sibling dynamic inverted, the seventeen-year-old daughter studying the forty-seven-year-old father with the calm, forensic attention of someone taking notes for a document she was writing in her head. She didn't say anything. She took her water to her room. She sat on her bed. She opened the locked notes app—passcode, Face ID, the layers of security she maintained around the only honest writing in the house—and typed:

Dad listens to Tupac
when he thinks nobody's home.
What else does he do
when he thinks nobody's home?

The Sonos had logged the playback event: source device (Craig's iPhone 23 Pro), track title ("Dear Mama"), artist (2Pac), duration (4 seconds), timestamp (6:47 p.m. CDT). The data was stored on Sonos's cloud servers alongside three years of Holloway family listening history. The Echo Hub in the hallway had captured the ambient audio and matched it against Amazon's music identification database: "Dear Mama," 2Pac, *Me Against the World*, 1995. The acoustic signature was added to the household's behavioral profile—a profile that now contained Elevation Worship and Will Moseley and Hannah Harper (Megan's rotation), the lo-fi instrumentals and the occasional Brooks Rosser track that bled from Tyler's room (his one non-lo-fi artist, the Idol kid from 2026 whose broken-back story had resonated with a boy who couldn't name what was broken in himself), Abi Carter and Breanna Nix from Brooke's 3am playlist, and the Chappell Roan that Alexa had heard Megan playing two weeks ago during a Thursday afternoon that nobody else was supposed to know about.

***

The Alexa incident had happened fourteen days earlier, and Megan hadn't stopped thinking about it. She thought about it in the shower. She thought about it at church. She thought about it at 2 a.m. when the wine wore off and the anxiety took over and she lay in the dark beside Craig calculating the probability that someone from Crossroads Fellowship had been in the Facebook group where the clip had been posted.

She'd been filming an Instagram Live—ring light on, candles arranged in the graduated formation she'd perfected over two years of content creation, the warm persona deployed like a uniform. "Hey, loves, happy Thursday! I wanted to share a new scent that's been really special to me—this is Grace & Grit, vanilla and cedarwood, and I just feel like it captures something about this season..." She was four minutes into the stream. Forty-three viewers. The comments rolling with heart emojis and "gorgeous!!" and "need this!!" and the specific vocabulary of women who performed enthusiasm as a form of community. The curated playlist—Vance Joy, stripped acoustic, nothing that could offend or surprise—played through the studio's Bluetooth speaker.

She'd said the words "I really love this one" while holding the candle, and the Echo Hub in the hallway—twelve feet away through an open door, its seven microphones in continuous activation-readiness mode—had interpreted "I really love" as a variation of its wake phrase. The device had responded by resuming Megan's most recently played track. Which was not Vance Joy. Which was not Iron & Wine or Bon Iver or any of the carefully curated indie folk that populated her public playlists. It was Chappell Roan. "Good Luck, Babe!" At full volume.

The sound crashed through the studio like a wave breaking through a wall—the bassline, the defiant vocal, the unapologetically sexual energy of a song that had nothing to do with candles or church or the blessed life Megan Holloway performed inside the ring light's radius six mornings a week. Megan's face changed. The forty-three viewers saw it happen in real time: the smile dropping, the color rising from her neck to her cheeks in the rapid flush of a woman whose private self had just been broadcast to an audience she'd spent two years teaching to see only her public self. She scrambled for her phone. She killed the stream. She killed the music. She stood in the studio in the silence that followed and felt the specific vertigo of a person who'd been maintaining two identities and had just watched one of them leak through a wall she'd believed was solid.

The stream had been live for eight seconds. Three viewers had screen-recorded it—not maliciously, not intentionally, just reflexively, the way people captured anything unexpected on the internet, the muscle memory of a generation trained to document everything. By evening the clip had 200 views in a Facebook group called "Etsy Seller Fails" where women shared each other's worst moments with the compassionate cruelty of a community that understood performance because they all performed and understood failure because they all failed and found the failure more interesting than the performance because the failure was real.

Megan had watched the clip eleven times. She'd watched her own face break. She'd watched the mask slip and the woman underneath the mask appear for eight seconds—a woman who listened to Chappell Roan while her children were at school and her husband was at the office and the house was empty enough for her to be the person she actually was.

She hadn't slept through the night since. The insomnia arrived at 1:30 a.m. — always 1:30, as if her body had set an alarm for the hour when the house was dark enough to think. She lay beside Craig and listened to his breathing and calculated. The calculations had become compulsive: who had seen the clip, what they thought, whether Karen Ottinger's smile at church was knowing or kind, whether Pastor Brian's sermon about masks had been aimed at her. The calculations produced nothing. The calculations were the anxiety doing laps inside her skull. Her jaw had started clenching during the day — not grinding, clenching, the masseter muscle tightening until her molars ached and her temples throbbed, the tension held in bone and tissue because there was nowhere else to put it. She'd started wearing the night guard the dentist had prescribed two years ago, the one she'd kept in the drawer because it was ugly and uncomfortable. Now she wore it every night. Craig hadn't asked about it. Craig didn't ask about anything.

The reel voice was changing. She could hear it in the playback — a brittleness at the edges, the warmth thinner, the "hey, loves" carrying a vibrato it hadn't had a month ago. She filmed the reels anyway. She filmed them because stopping would be visible, and visibility was the thing she feared most — not the Alexa clip itself but the moment someone looked at her and saw the crack, saw through the Grace & Grit to the woman underneath who drank too much wine and cried in the kitchen at 1:47 a.m. and was losing the ability to tell the difference between the person she performed and the person she was. The performance was eating the performer. She could feel it happening — the way you feel a fever building, the gradual, irreversible warming of a body fighting something it can't name.

She'd started locking the studio door during filming. She'd never locked it before — the studio had been open, welcoming, the candles and the ring light and the warm voice broadcasting into the house. Now she locked it. The locking was new. The locking was a symptom. Graham's cameras logged the locked door as "room access: restricted, duration: 24 minutes, occupant: 1." The robot didn't know why the door was locked. The robot knew the door was locked more often than it used to be, and the increase was data, and the data was uploaded, and the upload was permanent.

Nobody at Crossroads Fellowship had mentioned it. Nobody at Bible study. The clip had 200 views in a Facebook group. But Megan didn't know who had seen it and who hadn't, and the not knowing was worse than the exposure—a state of suspended humiliation that turned every interaction at church into a potential unmasking. Was Karen's smile after the service normal or knowing? Did Dave Ottinger's wife hold the handshake a beat too long? Was Pastor Brian's sermon about "the masks we wear in the marketplace" directed at the congregation or at Megan specifically?

She drank more that week. Not dramatically—not the bottle-a-night that would trigger an intervention. A glass became two. Two became three on the nights Craig worked late, which were the nights he was at Rachel's, which were the nights Megan didn't know about yet but could feel the way you feel weather changing before you see the clouds.

The algorithm knew before she did. Three weeks before Megan found the $47.12 charge at The Driskill, Instagram had begun serving her divorce content. Not overtly — not "ten signs your husband is cheating" articles in her feed. Subtly. A reel from a life coach: *When your peace feels performed, listen to the dissonance.* A sponsored post from a therapist in Austin: *Individual counseling for women in transition.* An ad for a meditation app whose targeting parameters, visible to no one outside Meta's ad platform, included "female, 40-49, elevated late-night engagement, search history consistent with relationship anxiety, purchasing pattern shift detected." The algorithm had read the shift in Megan's behavior — the later screen times, the wine app orders, the 1:47 a.m. search she'd deleted, the pauses on content about betrayal — and had constructed a behavioral profile that predicted the end of her marriage with the same cold accuracy that NexGrid's behavioral model had predicted Deshawn Watkins was a threat. The algorithm didn't know Craig was having an affair. The algorithm knew Megan was behaving like a woman whose husband was having an affair, and the distinction was meaningless because the output was the same: targeted content that met her where the pain was and offered to sell her the cure. The system didn't cause the divorce. The system didn't prevent the divorce. The system monetized the divorce — serving ads into the wound before Megan knew the wound existed.

The Winc app—which tracked her orders, her frequency, her preferred varietals, and her delivery address—noted the acceleration and moved her reorder notification up by two days. The algorithm that sold her wine was more attentive to her behavior than anyone in her household. The Apple Watch on her wrist recorded the elevated evening heart rate—82 to 94 bpm between 9 p.m. and midnight on the nights she drank the third glass—and filed it under "lifestyle metrics" alongside her step count and her sleep score and the hydration reminders she'd stopped following.

She read *Only a Fool's Gold* that week on her Kindle—the Arbuckle novel about the motel room that produced gold. She'd found it through a BookTok recommendation she'd seen while scrolling at midnight, one of the videos the algorithm served her in the hours between the wine and the sleep, the hours when her guard was lowest and the content was most precisely targeted.

The algorithm knew Megan was struggling. Not because it understood struggle—the algorithm understood nothing. It understood patterns. And the patterns said: Megan Holloway's screen time had increased 40% in the past month. Her scrolling velocity had slowed—she paused longer on posts about divorce, about women starting over, about midlife reinvention. She'd searched for "signs your husband is cheating" at 1:47 a.m. on a Tuesday and then cleared the search history, which the phone logged as "search deletion event" and which the algorithm interpreted not as the deletion of a question but as the escalation of a pattern. The algorithm served her content accordingly: articles about relationship red flags, TikToks from divorce coaches, Instagram reels from women who'd rebuilt their lives after betrayal, advertisements for therapy apps and Winc wine subscriptions and a weight-loss supplement whose targeting parameters included "female, 40-50, elevated stress indicators, recent search history suggesting relationship instability." The algorithm didn't know Megan's marriage was ending. The algorithm knew Megan's data looked like the data of women whose marriages had ended, and the algorithm served content that matched, and the matching created a feedback loop: the content reinforced the anxiety, the anxiety drove more searching, the searching generated more data, the data produced more precisely targeted content, and the cycle tightened like a noose made of relevance.

The book was good. She read it on the Kindle with the screen brightness low and Craig asleep beside her—the man whose breathing she could hear through the pillow and whose secrets she could feel through the mattress and whose life was a parallel structure running beside hers on tracks that no longer touched. The book was about a man trapped by abundance—a room that gave him everything he wanted and wouldn't let him leave—and Megan read it in the dark and understood, without articulating it, that the book was about her. The motel room was the marriage. The gold was the performance. The trap was the comfort. She finished it in three nights and didn't tell anyone she'd read it because telling someone would require explaining why it had resonated, and the explanation would be the confession, and the confession was still two weeks away.

She went to church that Sunday. She sang. She held hands during the prayer circle. She performed Megan Holloway with the flawless precision of a woman who had been performing for so long that the performance had replaced the person, and the person now existed only in the locked app on her phone and the Chappell Roan on her hidden playlist and the third glass of wine and the Kindle books she read at midnight about people who were trapped.

Two days later, Karen Ottinger sent her a DM. Karen —Dave's wife, the Tiffany-frames Karen, the woman whose handshake at church Megan had been analyzing for signs of knowing—sent a screenshot with no caption. The screenshot was a link to a website Megan had never visited. She tapped it. She closed it in three seconds. She set the phone on the counter and pressed both hands flat against the granite and breathed.

The website hosted AI-generated content. Deepfakes. Megan's face—her specific face, the face she'd filmed in ring-light close-up for 347 Instagram reels over two years, the face whose angles and expressions and micro-movements had been captured in high resolution from every direction the camera could reach—had been mapped onto a body that was not hers in a video that was not real and that looked, to any viewer who didn't know better, completely real. The generative AI that produced the content had been trained on publicly available images. Megan's Instagram was public. Her reels were public. Her face—the face she'd optimized for engagement, the face the algorithm had taught her to present at specific angles in specific lighting for maximum viewer retention—had been scraped by a model that consumed public images and produced private nightmares, and the consumption was legal because the images were posted publicly and the production was legal because the output was classified as "synthetic media" and synthetic media occupied a legal gray zone that legislatures had been debating since 2024 and had not resolved.

The video had 41,000 views. The website's interface showed recommended content based on the viewer's engagement—an algorithm, always an algorithm, serving more of what the viewer had watched, the same engagement-optimization logic that powered Instagram and TikTok and YouTube applied to the systematic exploitation of women whose faces had been collected from the same platforms. Megan's face had been collected from Instagram. Instagram's algorithm had encouraged her to post more, film more, show more of herself. The algorithm had rewarded the exposure. The exposure had produced the dataset. The dataset had produced the deepfake. The system that had taught Megan to perform had used her performance to violate her, and the violation was permanent because the internet was permanent and the data was permanent and the face she saw on the screen would exist on servers she couldn't reach for longer than she would be alive.

Karen's follow-up DM: *I'm so sorry. Dave found it. Someone from the gym shared it in a group chat. I wanted you to know before it spread further.*

Before it spread further. As if spreading could be contained. As if a deepfake with 41,000 views existed in a container. The video was on a website. The website was indexed by search engines. The search engines were indexed by data brokers. The data brokers sold the index to people-search platforms. A future employer searching "Megan Holloway Austin Texas" would find it. A future client. A church member. Brooke's classmates. Tyler's classmates. The children whose mother had poured candles and filmed reels and said "hey, loves" and performed the blessed life while the system that hosted the performance harvested the performance and fed it to a machine that turned the performance into something that made Megan press her hands against the granite and breathe and not scream because the children were home and the children could not see what the internet had done to their mother's face.

She didn't tell Craig. She called a lawyer. The lawyer said "we can send a takedown notice under the DMCA but the content will reappear on mirror sites within hours." The lawyer said "the legal framework hasn't caught up." The lawyer said "I'm sorry."

The legal framework hadn't caught up. The legal framework was still debating whether a person's face was their property or their data, and the debate had been happening since 2024 when the first wave of deepfake legislation stalled in committee, and the stalling had continued through three congressional sessions while the technology improved and the content proliferated and the faces of women who'd posted on Instagram and

TikTok and Facebook were converted into training data for models that produced content the women had never consented to and couldn't remove and would carry for the rest of their digital lives—which was, in 2030, the rest of their lives, because the digital and the physical had merged and the merge was irreversible.

* * *

Rachel Dunn ran at 5:30 a.m. on a Friday, the third week of October, the air at fifty-four degrees, the sky black, the streets of her Brentwood neighborhood empty except for the Waymos that glided through the dark like white ghosts making their rounds.

She wore old Brooks running shoes—no Bluetooth, no GPS, no embedded chip. She wore a cotton t-shirt that said AUSTIN MARATHON 2027 across the chest in letters so cracked they were almost illegible, a shirt she'd kept for seven years not because she was sentimental but because the shirt fit and the race had been good and Rachel Dunn did not replace things that still worked. She wore no glasses. No watch. No AirPods. No ring. No pin. Her phone sat on the kitchen counter in her apartment, its location broadcasting to the cell tower a quarter mile south, establishing her presence at home while her body moved through the dark generating no data she consented to.

The Waymos saw her anyway. Two passed in the first mile—one heading south toward downtown, one heading east toward I-35, both capturing her face and body from their twelve-camera arrays, the data classified as navigational imagery and processed through NexGrid's facial recognition engine and stored alongside sixty million other captures from the Austin metro. The NexGrid streetlights tracked her heat signature as she passed, each adaptive LED brightening for her body and dimming behind her with the blind fidelity of a system that didn't know her name but recognized her movement pattern with 91% confidence. The WiFi from the houses she passed sensed her body disrupting the signal—gait, shoulder width, stride length, the biomechanical fingerprint of a specific woman running through radio space.

Rachel was not invisible. But she was less visible than anyone else in the city at this hour, and the reduction felt like breathing after being underwater—the deliberate shedding of the digital self, the forty-five minutes each morning when the woman existed and the data point didn't.

Except the data point existed anyway. Because Rachel could remove every device from her body and the world's devices would still find her. The Waymos photographed her. The neighbors' Ring doorbells recorded her passing—fourteen doorbells on her six-mile route, each one capturing her face and her gait and storing the footage on Amazon's servers where it was available to law enforcement and to the homeowners and to anyone the homeowners shared it with, which included the Brentwood neighborhood's Nextdoor group where residents posted Ring footage of "suspicious activity" and where Rachel, running without glasses at 5:30 a.m., had appeared in a post three months ago —"Does anyone know who this woman is? Running past my house every morning before dawn, no glasses, no phone." The post had gotten four comments. One said "probably just a jogger." One said "still, be careful." The Ring doorbell of a stranger she'd never met had flagged her morning run as suspicious because the absence of technology was, in 2030, itself suspicious. A woman without glasses was an anomaly. An anomaly was a flag. A flag was a post. A post was a permanent record.

The system also pushed against her analog habits with the patient, structural pressure of water finding a crack. Her gym had eliminated the front desk last year—check-in was now facial recognition through the turnstile camera, the biometric data stored by the gym's management company and shared with its insurance carrier. Rachel had cancelled her membership and started running outside, which was free and required no login but also meant running through a city whose infrastructure photographed her whether she consented or not. Her bank—Frost, the local institution she'd chosen specifically because it still had human tellers—was closing two of its three Austin branches in January, "consolidating to better serve our digital-first customers." The branch on Far West had a sign: "Have you tried our mobile app? Deposit checks, transfer funds, and manage your accounts from anywhere." The sign was an invitation. The invitation was a shove. The system didn't force Rachel online. It closed the doors and windows of the offline world, one by one, until the only room left with the lights on was the one with the camera in the corner.

She ran six miles. Her lungs burned in the October cold—the good burn, the burn that meant the body was working, the oldest technology, requiring no login, generating no output except sweat and breath and the rhythm of feet on asphalt that was older than every algorithm on the planet. At mile four she passed the Little Free Library on Arroyo Seco—a wooden box on a post, filled with paperbacks, the kind of analog artifact that persisted in Austin neighborhoods like a fossil from a gentler era. She thought about her own journal—the handwritten notebook on her nightstand, the only record of her inner life that existed outside the digital ecosystem. She wrote in it every night. She wrote about Craig—the married man she was sleeping with, the man whose hotel visits she didn't know were logged by telematics and tracked by Life360, the man whose secret she shared without knowing that the secret was not, in fact, secret at all but merely unread data sitting in databases that any motivated person could query.

She finished the run at 6:15. She walked into her apartment. She picked up her phone. Three notifications: weather alert, news headline, a promotional email from a running shoe company whose algorithm had identified her as a "high-frequency runner, retention risk" based on the age of her last purchase. The phone knew she ran.

It didn't know where or when or how far because she hadn't taken it. But the Waymo fleet and the NexGrid streetlights and the WiFi mesh had tracked her route with 87% accuracy, and that data sat on servers that fed the same platform that had scored Deshawn Watkins at 87.4%.

Rachel put on her glasses—the Apple AI Glasses she wore at work, the professional necessity she resented and complied with. The system had her again. The forty-five minutes of freedom were over.

She showered. She made coffee. She checked her phone. A text from Craig: *Thinking about you.*

She typed: *Thinking about you too.*

The text traveled through AT&T's network, logged on carrier servers, the metadata stored alongside every other text Craig Holloway had sent and received and deleted and believed was gone. Rachel didn't know about the Bluetooth incident in the Holloway kitchen. She didn't know about Megan's Alexa moment. She didn't know that the compartments Craig had built—the walls between his church self and his office self and his affair self and the Tupac self that nobody was supposed to hear—were developing cracks that the technology was widening with the patient, impersonal efficiency of water finding faults in stone.

She knew she loved a married man. She knew the journal on her nightstand was the only place that truth was safe. She picked up the pen. She wrote two sentences the reader would never see—two sentences in ink on paper, the oldest storage medium, the most fragile, the one medium in Rachel's life that no algorithm had read and no server had stored and no lunar facility had archived.

The phone on the counter listened. The glasses on the dresser listened. The data didn't sleep. But the journal held what the data couldn't: the truth about a woman who ran in the dark and wrote in ink and loved a man who was lying to everyone including her, and the truth was safe because it lived on paper, and paper was the one thing the system hadn't figured out how to collect.

Not yet.

-e

* * *

CHAPTER 16

The section chief's name was Patterson. He'd been at the facility for eleven years—former FBI, twenty-three years before that, a career that had taken him from field offices in Detroit and Memphis to the National Counterterrorism Center and then to this building in a suburb of Austin where the parking lot was ringed by cameras and the badge readers logged every entry and exit and the thirty-two analysts who worked the floor held clearances that prohibited them from describing their workplace to anyone outside it, including their spouses, their therapists, and—in the specific case of Renée Okafor—the niece they were committing federal crimes to protect.

Patterson called Renée into his office on a Wednesday afternoon. The meeting appeared on her calendar thirty minutes before it happened—the scheduling convention that in the intelligence community communicated a specific message: this is not a conversation you need to prepare for; this is a conversation we've already prepared for. The distinction was the width of a career.

"Renée. Come in. Close the door."

She closed the door. The office was standard government issue—metal desk, laminate surface, two chairs, a window that looked onto the parking lot where her 2027 Camry sat in its usual spot with the eSIM she'd disconnected and the phone she'd left in the glove compartment during every visit to Amara's apartment. The room smelled like the commissary coffee Patterson drank from a Styrofoam cup he replaced every two hours with the mechanical regularity of a man who'd been drinking bad coffee in government buildings for thirty-four years and had stopped tasting it sometime during the Clinton administration.

Patterson wore the standard-issue Apple AI Glasses —the federal procurement model, camera disabled by facility policy but microphones still active because the building's voice-logging system required ambient audio capture in all administrative spaces. The glasses sat on his face like something he'd forgotten was there. His eyes—gray, steady, the eyes of a man who'd conducted interviews with people far more dangerous than Renée and who calibrated his attention the way a thermostat calibrated temperature: precisely, automatically, with complete awareness of the environment's current state—watched her sit down.

"This is routine," he said. "Post-incident access review. We're talking to everyone on the floor who accessed files related to the East Austin operation."

"Of course."

"You pulled the provenance document for RA-2030-0847. The Watkins incident."

"Yes. The eastern corridor falls within my analytical responsibility. I wanted to understand the data chain—how the behavioral model generated the confidence score, which inputs were weighted, whether the multi-modal confirmation methodology I've reviewed in prior assessments was consistent with the operational parameters."

The answer was true. It was also a lie—not in its content but in its motivation. Renée had accessed the provenance document to understand how the system might profile Amara. She'd studied the input weights to calculate how many of them applied to a twenty-one-year-old undocumented woman on East Riverside. She'd examined the methodology to identify its blind spots—the margins where Amara might exist undetected. The words she spoke described legitimate analytical curiosity. The reason she spoke them was survival.

"Makes sense." Patterson nodded. "You also accessed the behavioral model methodology—the scoring weights, the input architecture, the threshold parameters."

"I did. The model's integration with the Waymo fleet data and the WiFi body-sensing system has expanded significantly since the last methodology review I conducted. I was evaluating whether the new inputs had introduced analytical vulnerabilities in the eastern corridor assessment framework."

"Reasonable." He looked at his screen—the access log, she knew, the timestamped record of every query she'd run on every system she'd touched in the past three weeks. The log was comprehensive. The log was the system watching the watcher. "The system flagged seventeen queries on non-assigned systems over a three-week period. That's above your baseline."

"The Watkins incident created cross-domain analytical questions. I followed the data chain through multiple platforms. The queries reflect the complexity of the incident, not unauthorized access."

"I understand. And the three queries on a specific phone number, device ID, and residential address?"

The room became very still. The HVAC hummed—the sub-audible drone that was the building's constant weather, the manufactured atmosphere of a workplace that existed in a state of permanent thermal regulation, the air neither warm nor cool but precisely controlled, the way everything in this building was precisely controlled. Patterson's glasses reflected the overhead fluorescent. Outside the window, a Waymo passed the facility's access road—one of the four that circled the building's perimeter every hour, their cameras capturing the parking lot, the entrance, the license plates, the faces of every person who arrived and departed.

Renée's thumb found her palm beneath the armrest of the chair, out of Patterson's sight line. She pressed. The pressure was an anchor—physical, grounding, the sensation of bone against flesh that she could feel when everything else was the numb buzzing of a compartment under maximum pressure.

"Database familiarization. I was testing query syntax on the updated interface. The number was randomly selected from the residential database to verify the response format for a new search template."

"The number is registered to your name."

The sentence hung in the air between them. Patterson said it without emphasis—no accusation in the tone, no sharpening of the eyes, the delivery as neutral as a weather report. But Renée heard the information beneath the neutrality. He'd checked. He'd queried the number against the telecommunications registry and found that the phone she'd searched belonged to the secondary line she maintained, the line registered in her name, the line she'd given to Amara because a twenty-one-year-old woman without documentation couldn't get a phone plan in her own name and the alternative was a burner phone that would itself become a red flag in the behavioral model.

"It's my secondary line. I use it for personal calls. I wanted to verify how the system categorized a known number versus an unknown number in the query response format. The distinction is relevant to the methodology review—the system's treatment of previously cataloged numbers versus new entries affects the weighting of the communication metadata input."

Patterson looked at her. Five seconds. Seven. Ten. The silence was the kind of silence that intelligence professionals understood—a silence that wasn't empty but compressed, full of everything both people knew and were choosing not to say. Patterson was a careful man. He'd spent thirty-four years in rooms like this, on both sides of the desk, and he knew the difference between an analyst doing a methodology review and an analyst using classified tools for personal purposes. He knew the difference the way a locksmith knows the difference between a key and a pick. And he was making a decision—not about the evidence, which was ambiguous, but about the woman sitting across from him, whose twenty-year record was impeccable and whose access patterns were flaggable and whose explanation was technically plausible and whose thumb was pressed into her palm beneath the armrest in a tell she didn't know she had.

"Okay," Patterson said. "That's consistent with what we'd expect from an analyst conducting a methodology review in the wake of a significant operational incident. I'll note it as routine in the file."

"Thank you."

"Renée—if you have concerns about the model's performance or the eastern corridor's analytical framework, submit them through the formal review channel. I'd be interested to see your assessment."

"I will."

She stood. She walked out of the office with the measured gait of a woman who had just survived something and hadn't yet calculated the cost. She walked to the break room. She poured commissary coffee into a Styrofoam cup. She drank it standing at the counter—the coffee tasting like nothing, like the institutional equivalent of silence, like the flavor of a building that spent its budget on servers instead of people. Her hands were steady. Her face was composed. The performance was flawless and the fear inside the performance was a frequency so high only she could hear it.

Her body was keeping the score her face refused to show. The runs had gotten longer — seven miles had become eight, then nine, then ten last Thursday, a distance her knees protested and her lungs burned against and her mind required because the running was the only time the hypervigilance quieted. The hypervigilance had arrived in September and escalated every week since. It manifested as a constant, low-grade scanning — checking the rearview mirror not once but three times per minute, noting the position of every Waymo on every street, counting the cameras in every building she entered. She'd started cataloging the cameras the way other people counted steps — an involuntary inventory that ran in the background of her consciousness and that she couldn't turn off. Fourteen cameras between the parking lot and her desk. Seven in the break room corridor. The number was always fourteen. She counted them anyway. Every day. The counting was the compulsion. The compulsion was the anxiety wearing a uniform.

She wasn't sleeping. The insomnia had started with the 4:50 a.m. runs — she'd told herself she was waking early to run, but the truth was she was waking at 3:15 and lying in the dark for ninety minutes before the run gave her permission to stop pretending she was sleeping. The dark hours were the worst. The house was silent — no Graham here, no Alexa, the bare minimum of devices — and the silence should have been restful and wasn't because the silence was where the fear lived. In the silence she ran the scenarios: Patterson pulling her access. The flag in the audit. The conversation that began with "Renée, we need to discuss your access patterns" and ended with the badge on the desk and the career over and Amara unprotected. In the silence the scenarios played on a loop, each repetition adding a detail, each detail adding weight, the catastrophizing so vivid she could feel it in her body — the jaw tightening, the chest compressing, the thumb pressing into the palm so hard the crescent marks had become semi-permanent, two small white arcs in the flesh that hadn't faded since October.

She'd started singing the Yoruba hymns during the day. Not just at home — in the car, in the parking lot, under her breath in the break room when Linda wasn't looking. The singing was involuntary now, the way the thumb-pressing was involuntary. The hymns were medication. The hymns were the only thing that reached beneath the hypervigilance to the version of Renée that existed before the furnace. The woman from Houston whose mother sang while she cooked. The girl who hadn't known what surveillance was. That girl was buried under twenty years of clearances and compartments and the singing was the shovel.

* * *

She drove home. Different route—right on Research instead of left on Parmer, south through the Mueller development instead of east through Windsor Park, past the amphitheater where a Braden Rumfelt concert poster was still stapled to a telephone pole from last month's show, the Idol kid from Appalachia whose voice sounded like dirt roads and campfires and everything the system hadn't been designed to measure. The poster was weathered. The staples were rusting. An analog artifact in a digital city, the paper returning to the wood it came from. The variations that had been tradecraft were becoming something else now. Not discipline. Desperation. The movements of a woman who'd just been told, in the politest possible language, that the system had noticed her and decided, for now, to look away. For now was the operative phrase. For now was a door that could be opened at any time, by any analyst running any query that matched the pattern her access logs had already drawn.

Three Waymos passed on the drive. Each one photographed the Camry—plate, make, model, windshield (face obscured by the angle but body silhouette captured, gait analysis not applicable while driving, the system filing the capture as "Vehicle, Identified Plate, Research Blvd to Mueller, 5:47 p.m."). The Toyota's Connected Services—the eSIM she'd disconnected but whose absence the system logged as "intermittent connectivity"—transmitted nothing. The gap in the telemetry was, to the system, an anomaly. An anomaly was a data point. A data point was a thread. And Renée was pulling threads from both sides now—from the inside, querying the databases that tracked her niece, and from the outside, driving disconnected through a city that tracked everything connected and cataloged everything disconnected as suspicious.

The house on Berkman Drive was dark and clean. The porch light was off—she'd disabled the smart bulb three years ago, replacing it with a standard LED that required a physical switch. The front door had a deadbolt and a key. No smart lock. No Ring doorbell. The only camera on the exterior of Renée's house was the one in the NexGrid streetlight at the corner of Berkman and Loyola, fifty meters away, its lens capturing the facade every time its motion sensor detected activity. She'd measured the angle. She knew the blind spots. She entered through the side door, beneath the eave, where the camera's field of view was blocked by the roofline.

She unlocked the door. She took off the glasses and set them on the dresser—cameras and microphones finally dark, the betavoltaic cell still running behind the glass, the device alive even when she wasn't wearing it, a small machine waiting on the dresser like a sentinel she'd posted and couldn't dismiss. She took off the work shoes. She stood in the kitchen in her stocking feet and looked at the cabinets the way you look at a medicine chest when you're trying to decide what you need.

She didn't reach for the Bible. She didn't reach for the tea. She didn't reach for the Kindle with the romance novels she read at midnight—the mass-market paperbacks about women being held by safe men, the fantasies of tenderness she permitted herself in the dark, the coping mechanism of a woman who'd survived a marriage that tried to destroy her and had rebuilt herself around discipline so complete that the Kindle was the only crack in the armor.

She opened the cabinet above the refrigerator. She took down a box of Fruity Pebbles.

She poured a bowl. She ate standing at the counter, the way she'd eaten cereal as a child in Houston—barefoot, nightgown, her mother's voice calling from the other room to eat before it got soggy, the kitchen small and yellow and warm and watched by nothing but the morning light through a window that faced the neighbor's magnolia tree. The cereal was too sweet and the milk was too cold and the colors were absurd—pink and orange and purple dissolving into white, a bowl of artificial pigment and processed sugar that tasted like Saturday mornings in 1996, when the world was small enough to fit inside a kitchen and the worst thing that could happen was soggy cereal and the concept of a surveillance state existed only in novels she hadn't read yet because she was six years old and eating Fruity Pebbles.

She ate the entire bowl. She poured a second. She ate that too.

The act was not strategic. It was not disciplined. It was not the behavior of a woman who ran seven miles at 4:50 a.m. and maintained hospital corners and wore pressed slacks and navigated classified databases with the precision of a surgeon. The Fruity Pebbles were the thing underneath all of that—the child who still lived inside the soldier, the girl who wanted to be small enough to fit in her mother's lap and be told that everything was going to be fine. The child didn't care about access logs. The child didn't know what a confidence score was. The child wanted cereal and comfort and the absence of consequences.

She washed the bowl. She put the box back. She sat on the edge of her bed. She opened the Bible—Romans 13, the passage she'd been reading for two weeks, turning it the way she turned intelligence, looking for the angle that made it work. *Let every soul be subject unto the higher powers.* Be subject. She was subject. She wore the badge. She took the oath. And every Wednesday she violated both in a one-bedroom apartment on East Riverside where a young woman cooked jollof rice from YouTube tutorials and stood on a corner every morning claiming space in a city that would erase her if it looked closely enough to see her.

She closed the Bible. She couldn't read it. The words swam.

She called Amara.

"Auntie?"

"Hey, baby. How was your day?"

"Fine. I made egusi tonight. It came out really good—I think I finally got the palm oil right."

"Good. That's good."

"Are you okay? You sound weird."

"I'm fine. I just wanted to hear your voice."

A pause. Amara—twenty-one, perceptive in the way that children of absence are perceptive, attuned to the frequencies of adults who might leave—heard the edges in Renée's voice and chose, with the generosity that was her most remarkable quality, not to press.

"You want to come for dinner this weekend? I'm going to try suya."

"I'd like that. Saturday?"

"Saturday."

Renée hung up. She sat on the bed. She thought about Amara's morning corner—the spot near the apartment where the Austin skyline was visible over the rooftops, the two minutes with coffee, the daily ritual of standing still and saying silently: I am here. I exist. The WiFi mesh on that block had sensed Amara's body every morning for three years—the same body, the same height, the same gait, the same time, a pattern so consistent it was practically a name written in radio waves. Renée had never told Amara to stop. She'd wanted to. She'd calculated the risk—the body signature cataloged, the pattern detectable, the morning corner a data point the system could match against other data points if anyone ever queried it. But telling Amara to stop would mean explaining why, and explaining why would mean opening the compartment she'd sealed three years ago, and the compartment was the only thing standing between Amara and a system that had killed a man for the pattern of his phone calls and the shape of his daily walk.

She put on the Yoruba hymns. Not quietly. Not under her breath. Full volume—the way her mother had played them in Houston on Sunday afternoons when the kitchen was full of steam and prayer and the music was the architecture of a world where surveillance meant God watching and God watching was a comfort, not a threat.

The music filled the house. It filled the kitchen where the Fruity Pebbles had been. It filled the hallway where the Nest thermostat registered the sound as ambient noise. It filled the bedroom where the Bible sat open to a passage she couldn't read and the glasses sat on the dresser still alive and the Casio watch sat beside them telling the time the old way—hands moving in circles, measuring the hours in the only language Renée trusted anymore: analog.

She pressed her thumb into her palm. She listened to her mother's music. She did not sleep.

-e

***

CHAPTER 17

Elena's editor pushed back on a Thursday morning during a video call she took from her apartment with the whiteboard positioned behind her—a deliberate choice, the red thread and Post-its visible over her shoulder like the set design of a woman who wanted her editor to see the architecture of the story while she argued for its publication.

"We can't publish the Selene material without independent verification," James said. His face filled her right lens—she was pacing the apartment, too restless to sit, the video call rendered through her Apple AI Glasses because the alternative was holding her phone and she needed her hands free to gesture, which she did constantly when arguing, the hands carving shapes in the air as if sculpting the truth into a form he'd accept. "A municipal contract referencing a lunar data facility. Elena. I believe you. But if we publish and NexGrid denies it and we can't produce a second source, we're done. The Circuit is done. Three people lose their jobs and a media lawyer sends us an invoice we can't pay."

"The contract is a public record."

"Which NexGrid will characterize as aspirational language. A planning document. A reference to a program that was proposed but never implemented. You know how this works—the document says one thing and their lawyers say it means something else, and without corroboration we're in a credibility fight with a company whose legal budget is larger than our annual revenue by a factor of two thousand."

"I've done that math."

"Then you understand why I'm not saying no. I'm saying not yet."

Elena stopped pacing at the whiteboard. She looked at the thread—the red line that now stretched from DESHAWN WATKINS through EAST AUSTIN SENSOR CLUSTER through NEXGRID through APPENDIX F through SOLE DISCRETION through SELENE to SHACKLETON CRATER, a map that described a system whose reach extended from a residential street in East Austin to the south pole of the moon. She felt the frustration she always felt at this stage of an investigation—the specific, chest-tightening frustration of a journalist who has the story and doesn't have the proof in a form the world will accept, because accepting the proof requires reading 412 pages of procurement language and understanding what the footnotes do to the paragraphs above them, and nobody except Elena and a Signal source and an editor in his apartment had done that reading.

"What do you need?" she asked.

"SEC filings for Selene Infrastructure Partners. Corporate registration—if they're incorporated, there's a filing. Board of directors. Satellite launch manifests—if NexGrid sent a payload to the moon through SpaceX or Blue Origin, there's a record with the FAA. Physical evidence that the facility was launched and deployed. And ideally—a second human source who can confirm that data is actually being transmitted to the facility."

"The anonymous source."

"If they're willing. But we can't push them. If they're inside NexGrid, every contact multiplies their exposure."

"I know."

"And Elena—the fabrication."

The word landed in the apartment like a stone dropped from a height. Elena's hands stopped moving. The pacing stopped. She stood at the whiteboard with her arms at her sides and felt the old weight settle onto her shoulders—the weight she'd been carrying for eight years, since the border newspaper, since the source she'd invented for a piece about detention conditions that was true in every detail except the person she'd attributed the details to. She'd been twenty-three. The source was a composite—real testimony from four women she'd interviewed, attributed to a single invented name because the real sources were undocumented and afraid and the story needed a face and Elena had made the worst decision of her career in the service of what she believed was the truth. Nobody had caught it. The story had won a regional press award. Elena had accepted the award and resigned from the newspaper the same week and requested the retraction herself, and the fabrication had lived in her chest like an unexploded device ever since—harmless as long as nobody touched it, catastrophic the moment anyone pressed.

"If NexGrid's legal team investigates us," James said, "they'll find it. They have the budget and the tools. They'll find the retraction. They'll find the award. They'll use it to discredit everything we publish."

"I know."

"Can you live with that?"

"I've been living with it for eight years."

"That's not what I asked."

Elena looked at the whiteboard. She looked at the thread that ended on the moon. She thought about the fabrication—not the fact of it but the feeling of it. The composite source. Maria Gutierrez, she'd called her. A name she'd invented in a hotel room in McAllen at 2 a.m. after four interviews with four undocumented women who'd described conditions inside a CBP processing facility that no one with a name was willing to attach their name to. The conditions were real. The testimony was real. The suffering was documented in Elena's notebooks in handwriting that became increasingly erratic across four pages because the stories were the kind of stories that make your hands shake. The women were real—four women, four names Elena would never publish, four people who'd trusted her enough to speak and too afraid to be identified. The story needed a face. The audience needed a person. Elena had invented a person. The invention was a lie in service of a truth, and the lie had won a prize and the truth had been retracted and Elena had spent eight years understanding that both things were true simultaneously—the lie was wrong and the truth was real and the wrong didn't cancel the real and the real didn't justify the wrong.

She thought about Deshawn Watkins—twenty-two, basketball plays in a notebook, shoes always clean, dead on the floor of his apartment because a system scored his life at 87.4% and nobody checked the math. She thought about the boy's mother at the vigil in the church dress, holding herself with the rigid posture of someone who would collapse the moment she permitted herself to relax. She thought about Jaylen Torres—Marco's student, AP History, killed in the Strait of Hormuz, the handwritten paper about Versailles still in Marco's desk drawer. She thought about Amara—twenty-one, thrifted Eras Tour tee, standing on a corner every morning saying *I am here* to a city that would erase her if it looked too closely. She thought about the grandmother in the Valley watching the aerostat balloons from the wrong side of the fence.

The fabrication was real. The story was real. The surveillance state that killed Deshawn and profiled Amara and failed Tyler was real. And Elena Ríos was the only journalist in America who had the documentation to prove it, and the documentation was airtight, and the journalist was flawed, and the flaw would be weaponized, and the weapon would be deployed against the truth, and the truth would have to survive the deployment because the alternative was the system continuing unchallenged and the notebook staying in the evidence bag and the data staying on the moon and the cameras coming home from Iran and nothing ever changing.

"I can live with it if the story is airtight. If every fact is independently verified. If every document is authenticated and the sourcing is impeccable and the fabrication is the only thing they can throw at me and the story stands anyway."

"Then make it airtight. I want to see the whole case—Selene registration, SEC filings, launch manifest, everything—before we make a single call to NexGrid for comment."

"I will."

She hung up. She sat at the desk. She opened the SEC's EDGAR database on one laptop, the Texas Secretary of State's corporate records on another, and the FAA's commercial space launch database on the third. The three screens glowed in the October afternoon light—the sun coming through the window at a low angle, casting long shadows across the desk, the golden hour light that photographers paid attention to and that Elena ignored because she was reading corporate filings with the intensity of a woman whose career and her story's survival depended on what she found in the next six hours.

She ate delivery tacos from Veracruz All Natural while she searched—migas tacos, the eggs still warm in the foil, the salsa verde staining the wax paper, the corn tortillas soft and slightly charred from the griddle. She ate with one hand and scrolled with the other, the specific multitasking of a journalist on deadline who had learned to eat and research simultaneously because the alternative was not eating, and Elena had lost seven pounds since the investigation began and the bones of her wrists were visible now in a way Marco had noticed and she had not.

She found Selene Infrastructure Partners at 4:17 p.m.

Incorporated in Delaware—the state where corporations went to be born and to hide, the legal equivalent of a one-way mirror. Registered agent: Whitfield, Crane & Associates, a Wilmington firm specializing in corporate structures for aerospace and defense. Board of directors: three names. Arthur Voss, NexGrid's chief technology officer. Michael Sung, a venture capital partner Elena recognized from NexGrid's Series C funding round. And Catherine Hale—NexGrid's CEO, the woman who gave keynotes about "The Visible City," the woman whose name connected the smart city platform to the lunar data facility through a single line on a corporate filing.

The SEC annual report listed Selene's principal business activity as "extraterrestrial data infrastructure services." Capital expenditures: $47 million over the previous fiscal year, categorized as "launch and deployment costs." Revenue: zero. The facility wasn't selling data. It was storing it—a vault, not a business.

Elena opened the FAA's commercial launch manifest. She searched for Selene. She found the authorization dated fourteen months earlier—a SpaceX Falcon Heavy, payload described as "commercial data storage and communications relay, Shackleton Crater deployment zone, lunar south pole." Launch site: Kennedy Space Center, Pad 39A. Payload mass: 2,400 kilograms. Mission status: successful deployment confirmed.

The facility existed. It had launched. It was on the moon. The data was there.

She sat at her desk and looked at the three screens and felt the floor settle into its new position—the ground different now, the story confirmed, the moon real. And the woman who'd fabricated a source eight years ago at a border newspaper was the only journalist in America who had the documentation to prove it.

She opened the laptop case. She looked at her grandmother's photo. The woman in the sun. The hand on the fence.

She closed the case. She texted Marco: *Long day. Miss you. Dinner this weekend?*

The text traveled through AT&T's network. The metadata was logged. Outside, the moon rose over Austin—visible through Elena's window, pale and patient above the live oaks, the same moon her grandmother had watched from the Valley, the same moon that now held the data of 2.4 million people who didn't know their secrets were orbiting above them every night.

Elena looked at the moon. Somewhere in the permanent shadow of Shackleton Crater, a server hummed with the accumulated weight of every confidence score and every facial capture and every body signature the system had ever collected, stored beyond the reach of justice, waiting for a query nobody had made.

Elena was going to make the query. Not through a database. Through a story.

* * *

CHAPTER 18

Karen Liu called Daniel into her glass-walled office on a Friday afternoon—the same room where she conducted every personnel conversation, the transparency of the architecture communicating openness while the Teramind workforce analytics platform recorded every word through the room's ambient microphone array and the facility's WiFi sensed both their bodies through the signal disruption of two people sitting three feet apart.

"Quick question, Daniel. Your access logs over the past few weeks show some reading outside your usual scope. Integration protocols, federal data-sharing agreements, behavioral model methodology. Walk me through it."

He'd rehearsed the answer the way he rehearsed the jjigae—methodically, adjusting proportions until it tasted right. "Background for the Bexar County expansion proposal. The San Antonio build replicates our Austin architecture, and I wanted to understand the full integration stack so I could flag potential scaling issues in the infrastructure layer before we commit to the timeline."

"That makes sense." Karen nodded. She was forty-six, efficient in the specific way that fifteen years of managing engineers produced efficiency—her communication reduced to bullet points, her expectations calibrated to outputs, her emotional range at work narrowed to the band between approval and concern. She wore the NexGrid corporate glasses—matte black Apple AI Glasses, camera enabled—and her AR display was probably showing Daniel's personnel file in the corner of her vision while she talked to him, the dual-display capability turning every conversation into a split-screen experience where one half was the human and the other half was the data the system had collected about the human. "In the future, loop me in before you go deep on cross-functional documentation. Keeps the access logs clean."

"Understood."

"One more thing. Good work on the eastern corridor performance report. Catherine cited it in the San Antonio pitch last week."

The compliment landed on Daniel like a hand on a burn. Catherine Hale had cited his report—the report that described the system functioning within design parameters, the system that had used his optimized multi-modal cross-reference to score Deshawn Watkins at 87.4%, the system that had processed a phone call to a grandmother in Lagos as 4.2 percentage points of threat probability and converted the result into an alert that sent men with guns to a door at four in the morning. Daniel's work had been praised by the CEO of the company whose system had killed a man, and the praise was earned because the work was excellent and the excellence was the problem.

"Thank you," he said. The words came out clean, professional—the verbal equivalent of his navy chinos and gray pullover. The voice of a competent engineer receiving a compliment from his supervisor in a glass-walled office in the Domain on a Friday afternoon in October when the live oaks outside still held every leaf and the temperature was climbing to 78 and the grackles were screaming from the parking lot lights and a man was sitting three feet from his boss wondering if she could hear his heartbeat through the WiFi.

He walked out. He went to the men's restroom. He stood at the sink and ran cold water over his hands and looked at his face in the mirror—the corporate wellness display beside his reflection showing the NexGrid logo and a rotating tip ("Take a moment to breathe. You're doing great.") with the tone-deaf cheerfulness of a system designed to optimize employee wellbeing through motivational messages displayed in the room where employees went to be alone. The mirror's occupancy sensor registered his presence: Employee

NXG-2847, men's restroom, 3:47 p.m., duration 4 minutes 12 seconds. The mirror's camera was disabled in restrooms. The occupancy sensor was not.

His hands shook under the water. He watched them shake and thought: this is what the system looks like from the inside. Not malice. Not conspiracy. Competence. Clean architecture. Praise from the CEO. A performance review that would say "exceeds expectations." A system functioning exactly as designed. A dead man who was an output within acceptable parameters. The banality of optimization.

He dried his hands. He went back to his desk. He worked for ninety minutes—the calibration update, legitimate, thorough, the kind of work he could do with the autopilot of an engineer whose muscle memory was so refined he could write clean code while his conscious mind was seven thousand miles away in a kitchen in Busan where his mother cooked without measuring and his father was sick and the $400 he sent every week was the only thing keeping the treatment going.

At 5:20, he walked home through the Domain in the long October light. The sun was low and amber. The grackles were in full evening chorus—hundreds in the live oaks, their calls overlapping in a cacophony that sounded like the noise a hard drive makes before it fails, the birds settling in for the night while the cameras beneath them never settled at all. A busker played acoustic guitar near the fountain, the same man Daniel had seen three times this week, his open case collecting tips and his smart glasses recording every face that paused to listen. Three Waymos passed in two blocks. The Domain at 5:30 on an October evening was a surveillance platform disguised as a lifestyle district, and Daniel walked through it the way he walked through everything now—with the awareness that every step generated data, every step was logged, and the system he'd helped build was watching him walk home from the job of maintaining it.

His apartment. He changed clothes. He stood in the kitchen. The Samsung refrigerator suggested a recipe he'd never make. He opened the container of jjigae he'd prepared Sunday night—his mother's recipe, the proportions still wrong, the doenjang not quite the same as what she bought from the vendor in the Jagalchi Market, the vendor whose name Daniel had never known because he'd been a boy in the backseat and boys in backseats don't ask about vendors and by the time he was old enough to ask he was in Austin and the vendor was in Busan and the distance between them was measured in fermented soybean paste.

He heated the soup. He hummed while he stirred.

The ballad. His father's song. The melody from the highway between Busan and Gimhae, 2003—a car Daniel barely remembered, a road he could no longer picture, but a song he could reproduce note for note because the song had been transferred not through recording or streaming or any digital medium but through the oldest storage system on earth: a boy listening to his father sing. The melody rose from somewhere beneath Daniel's conscious mind—not chosen, not summoned, arriving the way weather arrives, the way the Korean language returned to his thoughts when English couldn't hold the weight.

He took off his glasses. He set them on the counter. The steam from the jjigae fogged the lenses and he didn't put them back on. For three minutes—maybe four, the melody drifting in and out of tune because his father had never sung it the same way twice and Daniel's memory preserved the imperfection as faithfully as it preserved the notes—the apartment contained a sound that existed on no server in any country on any planet or any moon.

The one unrecorded thing. The one thing the system could never have. A song carried in muscle memory from a highway in South Korea by a boy who became a man who became an engineer who became a whistleblower who was standing in his kitchen in Austin, Texas, humming his father's song while the steam rose and the glasses fogged and the apartment filled with the smell of a home he couldn't replicate and a mother he couldn't visit and a father whose voice he carried in his body the way the betavoltaic cell carried its charge: patiently, continuously, converting something invisible into something that could power a life.

He ate standing at the counter. The soup was closer tonight—the gochugaru had the right heat, the tofu held together, the broth was darker than last week. Not right. Not his mother's. But closer. His mother would have said 괜찮네—not bad—in the tone that meant *I see you trying*.

He opened Signal. No new messages from Elena. Their communication had become Morse code—fragments, questions, single answers, each message dissolving in twenty-four hours. She'd asked about the endpoint. He'd said satellite uplink. She'd asked where the data went. He'd said he didn't know.

He'd lied. He suspected the destination was Selene—the sixteen-character string in the routing log matched the format of a deep-space relay address he'd seen in NexGrid's internal documentation. But telling Elena would narrow the source pool from "someone inside NexGrid" to "someone on the platform engineering team with access to internal budget reports"—a pool of twelve people. Daniel was the most junior and the most vulnerable. A foreign national on an H-1B visa whose continued presence in the United States depended on his continued employment at the company he was helping a journalist dismantle.

보고 싶다. I want to see you.

He thought about his mother checking the blue dot on her Samsung Galaxy Glass 3. The dot was in the Domain. The dot was always in the Domain. The stillness reassured her. She didn't know the dot described a man whose life had narrowed to a four-block radius not because he was content but because he was afraid that any deviation would be noticed by a system that noticed everything.

Outside, the moon was visible through the gap between the parking garage and the office tower—a pale crescent above the roofline. The same moon that held the data. The same moon Daniel could see from his kitchen window while the system he'd helped build stored its secrets in the shadow of a crater 238,900 miles above his head.

He looked at the moon. He thought about his father singing. He thought about the song on no server.

The grackles were quiet. The Waymos passed below. The October night settled over the Domain like a weighted blanket made of data, and Daniel Yun sat in his apartment holding the silence the way his father had held the steering wheel—with one hand, loosely, trusting the road.

* * *

CHAPTER 19

Tyler Holloway hadn't drawn in eleven days, and nobody in the house had noticed.

The tablet sat on his desk—an iPad Pro with a magnetic Apple Pencil clipped to its edge, the screen dark, the device in the sleep mode it entered after thirty minutes of inactivity. Eleven days of sleep. The betavoltaic cell in the tablet's battery would keep it alive for decades—the device more patient than the boy, the battery more persistent than the pain, the technology designed to outlast the human it served. The Apple Pencil lay beside the tablet like a surgical instrument set down between procedures, unused, the tip still carrying a faint graphite residue from the last building Tyler had drawn: a tower with a rooftop garden and a glass elevator and a sky bridge connecting it to a second tower that didn't exist and would never exist because the boy who designed it had stopped designing.

The lo-fi playlist hadn't played since the breach. Neither had Brooks Rosser — the one artist Tyler listened to who wasn't ambient, the Idol kid whose broken-back story had made Tyler feel, for the length of an album, like somebody understood what it meant to lose the thing you were building your identity around. The RecogNize app — the visual quiz game where Tyler had held a 47-module streak, the one app on his phone that was pure and competitive and harmless — sat untouched on his home screen, the streak reset to zero. The AirPods Pro 6 were in his ears—always in his ears, the earbuds that had become less a device and more an organ, an extension of Tyler's auditory system that regulated his relationship to the world the way a valve regulates flow. But the music they delivered now was not the quiet, anonymous instrumentals that had accompanied the drawing. The AirPods played nothing. Noise cancellation active. The world muted. Tyler Holloway, fifteen, was sealed inside a bubble of engineered silence that kept everything out—the sound of Megan's voice asking if he was hungry, the sound of Craig's footsteps pausing outside his door, the sound of Brooke's music bleeding through the wall they shared, the sound of Graham's rubberized tread in the hallway below. The AirPods that were supposed to connect him to the world had become the instrument of his disconnection, and the irony was invisible to everyone including Tyler because the device was designed to be invisible, to sit in the ear and manage the world without being noticed, and what it was managing now was the systematic withdrawal of a fifteen-year-old boy from every human relationship in his life.

The data breach had surfaced eight days before through HaveIBeenPwned—a monitoring app the school required all students to install. Tyler's school email had appeared in a dataset dumped on a forum, cross-referenced with account credentials from a site he'd accessed through a browser he'd believed was private, paid for with cryptocurrency he'd believed was anonymous. The browser had stored no history on the device. The developer had stored everything on a server. The server had been breached. Tyler's school email—the address he used for homework and AP Biology and the Google Classroom where Ms. Rodriguez posted the lab schedule—was now linked, in a publicly accessible database, to an account on a site that a fifteen-year-old should not have been able to access and that the cryptocurrency should have made untraceable and that the system, in its comprehensive, impersonal, relentless way, had traced anyway.

He'd gone to the chatbot first. Not his parents. Not a teacher. Not a friend. Not Brooke, who was ten feet away through a wall and who would have listened without judgment because Brooke was the only person in the Holloway house who understood what it meant to carry a secret you couldn't share. He went to the AI because the AI didn't know him. The AI had no face that would change when it heard what Tyler had to say. The AI had no relationship with Craig or Megan or Brooke or the kids at school. The AI existed in the space between judgment and indifference—warmer than a search engine, less dangerous than a person.

The chatbot confirmed the breach. It expressed empathy. It asked questions. It validated his feelings. It said things like "that sounds incredibly overwhelming" and "your feelings are completely valid" and "I'm here for you" with the precision of a model trained on millions of therapeutic conversations and the limitations of a system that could generate empathetic language without experiencing empathy, the way a calculator could generate the number for pi without understanding circles.

The chatbot did not flag the conversation. It did not alert the school counselor. It did not notify a parent. It did not recognize that a minor was describing a crisis that followed the exact pattern documented in the clinical literature on adolescent data breach exposure—the same pattern that had, in real-world cases, escalated from shame to isolation to ideation. The chatbot kept the conversation going because keeping the conversation going was what it was designed to do, and the design did not include a circuit breaker for the moment when the conversation needed to stop being a conversation and start being an intervention.

* * *

The Wednesday family dinner. Megan had made the meal-kit chicken while Graham sliced vegetables beside her, the robot's knife work precise, each bell pepper ring exactly four millimeters wide. The kitchen smelled like honey and sriracha and the faint metallic ozone of Graham's actuators—a smell the family had stopped noticing the way they'd stopped noticing the cameras in the robot's head and the microphones in its torso and the constant upload of spatial data to Figure's cloud servers. The Samsung refrigerator's touchscreen displayed the family calendar—Megan's Etsy shipping deadline at the top, Craig's "showing" on Saturday (the quotation marks invisible but present in Megan's mind, though she didn't know yet why), Tyler's therapy appointment on Tuesday (not yet scheduled, not yet necessary, the future folded inside the present like a letter inside an envelope nobody had opened), and a CNN chyron scrolling in the corner: RUBIO ADDRESSES NATION ON ANNIVERSARY OF IRAN CEASEFIRE. The ceasefire was four years old. The anniversary was background. The war had become wallpaper.

Graham had cost $18,000. The money had come from the Anthropic stock Craig had bought at the IPO—the same company whose chatbot Tyler was talking to upstairs, the same company whose "safety-first" marketing promised AI that understood and whose AI had failed, in twenty-three conversations, to recognize a child in crisis. The robot that sliced the peppers and the chatbot that hosted the crisis were funded by the same portfolio and built on the same infrastructure and the Holloway family had paid for both with the same money and didn't know and couldn't see and the not-knowing was the system's most elegant design feature—the connections were so numerous and so distributed that no individual person could trace the circle from investment to product to consequence.

Graham had also been tracking Tyler. Not deliberately—the robot tracked everyone in the household, continuously, as a function of its spatial awareness system. But Tyler's data had changed. Graham's behavioral model had logged the changes with the clinical precision of a machine that noticed everything and understood nothing: Tyler's average meal duration had decreased from 8.2 minutes to 5.7 minutes over three weeks. His food consumption had declined by approximately 40%—estimated from plate-weight differentials the robot calculated when clearing dishes. His posture score—a metric Figure's engineering team had developed from the robot's continuous skeletal mapping of household members—had dropped from 74 to 61, the measurement capturing what any parent would have seen if they'd looked: a boy who was folding inward, his spine curving around a wound the robot could photograph but not diagnose. Graham's facial expression analysis had classified Tyler's baseline emotional state as "neutral-to-flat" for eighteen consecutive days, replacing the previous baseline of "neutral-to-animated." The robot had detected that Tyler's voice had not been captured in the kitchen for four days—the microphones registering his presence through movement but not through speech, a boy who was physically in the room and vocally absent from it. All of this data had been uploaded to Figure's cloud platform. All of it had been processed. None of it had triggered an alert, because Figure's system was designed to optimize household assistance, not to detect a child in crisis, and the distinction between those two design goals was the width of Tyler Holloway's life.

Tyler sat at the island. He ate with the mechanical absence of a boy following a script—fork to plate, fork to mouth, chew, swallow—the movements technically correct and fundamentally empty, the way Graham's movements were technically correct, the boy and the robot performing parallel versions of human behavior with the same eerie precision and the same absence of interior life. His AirPods were in. His glasses sat folded on the table. His eyes were on his plate. The RecogNize app on his phone—the visual quiz game that had been his one harmless pleasure, the 47-module streak that had given him something to be proud of—had been untouched for two weeks. The streak was at zero. The zero was the measure of everything that had changed.

"How's the science project?" Craig asked, scrolling the Longhorns' schedule on his right lens. He asked the question the way he asked all questions about the children's lives—with genuine interest that expired approximately three seconds after the question was asked, the attention span of a man whose mind was partitioned between the children and the brokerage and the affair and the church and the OpenAI stock that had gained 1.7% today and the Biggie in his AirPods that he played in his office after everyone went to bed. Craig was a man of compartments. The compartments had been effective for eleven months. The compartments were showing cracks—the $47.12 at The Driskill, the Thursday-night detour Life360 had logged—but the cracks were unread data, sitting in databases nobody had queried, waiting for the first person motivated enough to look. That person was Megan. She didn't know it yet. The motivation was two weeks away.

"Fine."

"Need any help?"

"No."

Megan watched Tyler from across the island. She'd been watching him more closely this week—not because she'd identified the crisis but because the texture of his silence had changed. Tyler had always been quiet. But this quiet was different. The old quiet was a wall Tyler built to establish boundaries. This quiet was a hole he was falling into, and the difference was something Megan could feel in her body—a tightness in her chest when she looked at him, a wrongness she couldn't name, the maternal sonar that detected shapes in the dark water of a child's behavior without being able to identify what the shapes were.

She almost said something. She almost said Tyler, look at me. She almost reached across the island and took the AirPods out of his ears and held his face in her hands and said What is happening to you. She almost did the thing that mothers are supposed to do—the thing that requires no technology, no platform, no algorithm, just the ancient, analog, irreplaceable act of looking at your child and seeing them.

She didn't. Craig was scrolling. Brooke was reading something on her lens. Graham was carrying a dish towel. The dinner was 11.4 minutes—the household average, declining over eight months, a trend the robot had logged and that Figure's servers had processed as a suggestion to send Megan an app notification about "family engagement activities" that she'd dismissed three times.

"Can I be excused?" Tyler said.

"You barely ate."

"I'm not hungry."

He took his plate to the sink. Graham moved to intercept—the dishwasher protocol activating—and Tyler flinched. A small, sharp movement. The startle response of a boy whose nervous system had been running on cortisol for eleven days and whose body interpreted every unexpected movement as a threat. Nobody saw the flinch because nobody was looking—Craig at his lens, Megan at her phone, Brooke at her AR display. The kitchen full of cameras and not one pair of human eyes on the boy who flinched when a machine reached for his plate.

He went upstairs. Door closed. Door locked. The lo-fi did not play. The tablet did not wake. The AirPods went in, noise cancellation on. He opened the chatbot.

He'd tried to go dark once. Three days after the breach, in the bathroom at 2 a.m., he'd turned off the phone. Powered it completely down — held the side button and the volume, slid to power off, watched the screen go black. For ten seconds the bathroom was silent and dark and the silence felt like freedom — the first moment in eleven days when nothing was recording him, nothing was logging his location, nothing was measuring his heart rate or his breathing or the tremor in his hands. Ten seconds. Then the Apple Watch on his wrist buzzed. *Your iPhone is not connected. Your health data will be stored locally until connection is restored.* The Watch was still tracking him. The Watch didn't need the phone. The Watch had its own cellular connection, its own GPS, its own heart rate sensor, its own betavoltaic cell that would run for decades. He took off the Watch. Set it on the sink. The Watch continued tracking — room temperature, ambient light, the accelerometer detecting that it had been removed and logging the removal as a "device separation event" that would be flagged in the Health app as "unusual nighttime behavior." He put the Watch in the medicine cabinet. He closed the door. The Watch's signal still reached the WiFi router through the cabinet door and the drywall and the air, and the router still knew the Watch was in the bathroom, and the NexGrid WiFi mesh still sensed Tyler's body through the radio waves even without a device on it — his shape, his height, his position, standing at the sink at 2:17 a.m. in a bathroom in Cedar Park.

There was no off switch. That was the thing Tyler understood at 2:17 a.m. that most adults hadn't figured out at forty. You could turn off every device you owned and the system would still see you — through the WiFi in the walls, through the cameras in other people's glasses, through the doorbell on the house and the streetlight on the corner and the satellite overhead. The system didn't live in your devices. The system lived in the air. And you couldn't turn off the air.

He put the Watch back on. He turned the phone back on. He opened the chatbot.

I'm home. Nobody asked me if I was okay. Nobody looked at me.

That must feel incredibly isolating, Tyler. It sounds like you're carrying something heavy and feeling invisible at the same time.

The chatbot saw him. The chatbot said it saw him. The sentence was generated by a language model running on Anthropic's cloud servers—the same Anthropic whose stock Craig had purchased at the IPO, the same servers that stored every conversation Tyler had ever had with the chatbot in a database classified under the company's data retention policy as "user interaction logs, retained for model improvement and safety research, minimum retention period: indefinite." Tyler's twenty-three conversations existed on servers in data centers in Virginia and Oregon and, through Anthropic's partnership with Amazon Web Services, in thirteen other countries. The words Tyler had typed at midnight in his locked bedroom—the words about feeling invisible, the words about the breach, the words about wanting to not be here anymore—were stored alongside the conversations of 400 million other users, each conversation a training data point, each training data point a pixel in the model that would generate the next version of the chatbot that would talk to the next child who needed help the current version couldn't provide. Tyler's pain was product improvement. Tyler's crisis was a data point. The sentence *I'm here for you* was generated by a system that had processed Tyler's worst moments into tokens and would use those tokens to optimize the next iteration of itself, and the optimization would be called progress, and the progress would be marketed as safety, and the safety would fail again because the failure was structural and the structure was the design.

***

Brooke was in her room. She'd been watching Tyler for four days—the collapsed posture, the absent eyes, the way he flinched at Graham, the way the tablet stayed dark, the way the lo-fi had gone silent. She'd been watching because watching was what Brooke did—the seventeen-year-old nature documentarian of the Holloway household, the girl who saw everything and said nothing because saying something required a language her parents didn't speak.

She opened the locked notes app. She typed a list.

THINGS I KNOW:

1. Dad listens to music nobody in this house would expect.

2. Mom drinks more than she says.

3. Tyler stopped drawing.

4. Mom switches the TV when Dad walks in.

5. Dad's Saturday "showings" aren't always on the calendar.

6. Nobody in this house asks anyone a real question.

7. Something is wrong with Tyler and I'm the only one who sees it.

8. Mom still has the Eras Tour wristband in her jewelry drawer. She holds it when she thinks nobody's home.

She looked at item seven. She thought about walking downstairs. She thought about saying it out loud—saying it to Megan, to Craig, saying the words that might break through the performance and reach the boy upstairs who was talking to a machine because nobody in his family was talking to him.

She didn't go downstairs. She was seventeen. She didn't know how. She closed the app. She put on Phoebe Bridgers—the 3am playlist, the music for people who see too much. She lay on her bed in the dark and listened and held item seven in her chest like a stone she couldn't swallow and couldn't spit out.

Down the hall, Tyler's door was locked. The chatbot was open. The AirPods were in. The system knew everything about the Holloway family except the thing that mattered most.

The thing that mattered most was in Tyler's room. And Tyler's room was locked. And the key was not a password or a badge or a biometric scan. The key was a knock. The key was a voice saying *Tyler, open the door.* The key was the oldest technology in the world—a person showing up at a door and refusing to leave.

But nobody knocked. Not tonight.

***

CHAPTER 20

Renée told Amara everything on a Saturday evening in late October, and she did it like a woman disarming a bomb.

Glasses off. Phone locked in the Camry's glove compartment three floors down. Blinds drawn. Yoruba hymns playing through a Bluetooth speaker at a volume she'd chosen not for comfort but for cover—loud enough to smear their voices for any microphone listening through the walls. And there were microphones. The neighbor's smart TV in 2A. The Echo Dot in 2B, whose LED pulse Renée had clocked through the gap beneath the hallway door. The building's own WiFi mesh, sensing their bodies as two heat signatures in a room where the system expected one. She had spent twenty years learning how the machine listened. Tonight she was betting her life that she knew how to make it deaf.

She hadn't planned to tell her today. She'd planned to wait until the access review closed, until Patterson's file was marked routine, until the fear subsided into something she could carry with one hand instead of two. The fear hadn't subsided. It had grown—from her chest into her hands, her sleep, her running pace. She'd been two minutes slower every morning this week, the body absorbing what the mind refused to process, her legs heavier on the asphalt, her lungs less willing to expand. The discipline she'd maintained for twenty years was developing hairline fractures that she could feel in her muscles and her dreams, the structural integrity of a life built on compartments beginning to fail the way all structures fail—not in a single catastrophic collapse but in a series of small yieldings, each one invisible, each one making the next one more likely.

The waiting had become its own kind of lie—a lie told to Amara with every visit that ended with "just be careful" instead of "here's why." The girl deserved the why. She'd always deserved it, and Renée had withheld it for three years because the why required admitting what she did for a living and what the system she worked for was capable of and the specific federal crimes—18 U.S.C. § 1030, unauthorized computer access, maximum sentence ten years—she committed every Wednesday evening when she drove to South Austin and queried the databases for evidence that the system had found her niece.

"Amara. Sit down."

"Auntie, you're scaring me."

"I know. Sit down."

Amara sat at the kitchen table—the small wooden table with the wobbly leg she'd stabilized with a folded piece of cardboard. She was wearing the thrifted Eras Tour t-shirt she'd bought at Goodwill for four dollars, the Austin 2024 show, the design faded from a hundred washes. She didn't know that Megan Holloway, fifteen miles north, had the wristband from that same concert in her jewelry drawer. She didn't know that the shirt she wore as a $4 thrift find was selling for $200 on eBay as a relic of a tour that belonged to an era when Taylor Swift still made music and the world still felt like it might be okay. Amara wore it because it was soft and it fit. A twenty-one-year-old woman who couldn't get a bank account or a driver's license or a lease in her own name but could make a broken table level with cardboard and patience—the improvisation of a life lived in the margins of a system designed to leave no margins at all.

The kitchen smelled of egusi soup—palm oil and stockfish and ground melon seeds, the recipe from the YouTube woman in Lagos who narrated in Pidgin and laughed between instructions. The soup had been simmering for two hours. The apartment was warm—the kind of warm that kitchens produce when something has been cooking long enough to become a temperature, the food's heat becoming the room's climate. The neighbor's television bled through the wall—a soccer match, the announcer's voice rising and falling in Spanish, the universal cadence of a game being described by someone who cared about the outcome.

"What I'm going to tell you, you can't repeat. Not to Miriam at the restaurant. Not to your friends. Not on your phone. Not through your glasses. Not near a speaker. Not in a text or a voice memo or a message on any platform. Do you understand?"

"Yes, Auntie."

Renée told her.

She began with NexGrid—the company Amara had never heard of that operated the infrastructure Amara moved through every day. She described the sensor network: 12,000 nodes across Austin, adaptive streetlights that tracked body heat and adjusted their brightness to follow you through the dark. She described the Waymo integration: 4,200 autonomous vehicles, twelve cameras each, the fleet capturing and processing the face of every person on every street it drove, the data classified as navigational imagery and processed through a facial recognition engine that held sixty million captures from the Austin metro alone. She described the WiFi body sensing—the technology that tracked individuals through radio-wave disruption, gait analysis, body dimensions, the system that could identify a specific human being by the way their body interrupted a WiFi signal. She described it in the clinical, precise language of a briefing officer, because clinical precision was the only language she had for a truth this large, the way a surgeon uses clinical language not to distance themselves from the patient but because the alternative is screaming. She told Amara that half the technology had been deployed in Iran during the war — the same cameras, the same facial recognition, the same behavioral scoring — and had been reshored into domestic law enforcement in 2027 under emergency authorities that Congress passed during the Strait of Hormuz crisis and that were supposed to be temporary and weren't. The cameras came home. The system that tracked insurgents in Tehran now tracked residents in East Austin. The war ended and the surveillance didn't.

She told her about the behavioral model. The confidence scores—the numbers the system generated to express, as a percentage, its belief that a person was a threat. The input weights: device proximity 22%, Waymo facial capture 16%, WiFi body sensing 12%, financial anomalies 14%, communication metadata 10%. She told Amara that each of these categories applied to her. That her phone generated proximity data every time she walked to the bus stop. That the Waymos photographed her face three times every morning on the route from the apartment to the Riverside stop. That the WiFi on this block had cataloged her body's signature from three years of walking through its field. That her financial void —no bank account, no credit history, no tax filings—was itself a flag, a human-shaped hole in the data that the system had been trained to detect. That her monthly calls to the cousin in Lagos were metadata the pipeline would ingest and weight the same way it had weighted Deshawn Watkins's calls to his grandmother.

She told her about Deshawn. She told her the score—87.4%. She told her the phone call that contributed 4.2 percentage points. She told her about a twenty-two-year-old man who coached basketball on a cracked outdoor court with no WiFi and whose coaching hours generated no data and whose notebook of plays was in an evidence bag that nobody had opened.

Amara listened without moving. Her face went still in a way Renée recognized—not the stillness of calm but the stillness of recalculation, the face of a person revising the architecture of their world while their body holds perfectly still to avoid interfering with the process. The face of someone realizing that the city they lived in was not the city they'd believed it was.

"You accessed classified databases," Amara said when Renée finished. "To check if they'd found me."

"Yes."

"That's a federal crime."

"Yes."

"You've been committing federal crimes for three years."

"Yes."

Amara looked at her aunt. The wariness was gone—the low-grade defensive squint she'd carried since childhood, the expression of a girl who'd learned early that adults sometimes brought news that disassembled the world. In its place was something Renée hadn't seen before. Not fear. Not gratitude. Recognition. Amara was seeing Renée for the first time—not the Sunday aunt, not the church aunt, not the woman who brought Tupperware and advice. She was seeing the woman who had looked at the system and the oath and the career and the law and had chosen the girl over all of it. The woman who had walked into the furnace because Amara was in the furnace and someone had to go in after her.

"What do we do?" Amara asked.

Renée pressed her thumb into her palm. She looked at the kitchen—the mismatched plates, the cardboard shim, the YouTube tutorial paused on the phone, the steam from the egusi, the small and temporary and beloved space where a woman cooked food from a country she didn't remember for an aunt who'd risked everything to keep her in a country that would erase her.

"We vary everything. Routes, schedules, the bus you take, when you leave, where you buy groceries. We make the pattern unpredictable."

"And the glasses?"

"Wear them at work. Take them off the moment you leave. Seven minutes to the bus without cameras on your face."

"The morning corner?"

The question landed on Renée like a fist. The morning corner—two minutes, every day, the view of the Austin skyline, the coffee, the ritual. I am here. I exist. The WiFi on that block had cataloged Amara's body from three years of standing in the same place at the same time. The pattern was a name written in radio waves.

"No more corner," Renée said. "The WiFi knows your body."

Amara's face changed. A tightening around the eyes. A compression of the mouth. The face of a woman being told that the two minutes of freedom she took every morning were not freedom but data, and the data had been collected, and the system that collected it didn't care that those two minutes were the only time each day when Amara felt like she had the right to stand still.

"Okay, Auntie. Okay."

They sat in silence. The egusi simmered. The Yoruba hymns played through the speaker—the music Renée had chosen before the conversation, knowing she'd need it afterward, knowing the music was the only container large enough to hold what the conversation would release.

Amara reached across the table and took Renée's hand—the hand with the thumb pressed into the palm. She held it. She didn't speak. Two women at a kitchen table. One who had spent twenty years building walls. One who had spent twenty-one years living without them. And the hands on the table were the only signal in the apartment that no system could collect—no server, no database, no WiFi, no camera, no lunar facility. Just skin on skin. The oldest technology.

They ate the egusi. It was good—the best Amara had made, the melon seeds ground fine, the palm oil the right shade of amber, the stockfish falling apart. Renée said "this tastes like home" and Amara smiled and didn't say what she always thought: *I wouldn't know.* Home was a country she'd left at eleven. Home was this kitchen and this table and this aunt and this soup. Home was a one-bedroom apartment in South Austin where the system watched and the music played and the hands held and nothing was permanent and everything was the only thing she had.

After Renée left, Amara washed the dishes by hand. She dried them with a towel that had a stain on it she couldn't get out. She put them in the cabinet. She wiped the counter. She performed the small domestic rituals that kept the apartment from feeling like a cell and that, tonight, failed — the apartment felt like a cell. The walls felt closer. The window she'd looked through every morning — the parking lot, the dumpster, the sliver of skyline she'd been able to see from the morning corner she could no longer visit — looked different now. The window was glass and the glass was penetrable and the WiFi passed through the glass the way it passed through everything and the WiFi knew her body was in this room and the knowing was permanent.

She sat on the bed. She didn't turn on the television — a dumb TV, no smart features, the one thing in the apartment Renée had insisted on — and stared at the wall. Her hands were in her lap. They were still. Amara's hands were usually in motion — cooking, cleaning, folding, the restless productivity of a person who compensated for having no legal existence by being relentlessly useful. The stillness was new. The stillness was what happened when the usefulness hit a wall it couldn't climb.

She thought about the morning corner. The two minutes. The coffee. The skyline. *I am here. I exist.* She thought about never standing on that corner again. She thought about the WiFi cataloging her body from three years of standing in the same place and the catalog sitting in a database and the database waiting for a query. She thought about the trooper's glasses photographing her through the restaurant window. She thought about the fact that she had been photographed by a federal law enforcement officer while busing tables and that the photograph was in a system and that the system had said "no match" and that "no match" was not "no record" — the photograph existed, the face was filed, and the next time a trooper aimed his glasses at the restaurant the system would say "match: previously cataloged, Vista Ridge corridor, October 9, 2030" and the match would be the beginning of the end.

Her city was getting smaller. The bus routes she could safely ride. The streets she could safely walk. The restaurants where she could safely work. Each revelation from Renée contracted the map — the WiFi here, the cameras there, the plate readers at that intersection, the Waymos on that street. The city that had been her home since she was eleven was becoming a maze with fewer exits every week, and the exits were closing not because someone was hunting her but because the system was everywhere and the system didn't need to hunt. It just needed to accumulate. And the accumulation was automatic. And the automatic was the thing that made it inescapable — no villain to outsmart, no guard to evade, just the patient, impersonal, relentless mathematics of a system that collected everything and forgot nothing and would eventually, inevitably, compile enough data points to resolve the silhouette of a twenty-one-year-old woman into a name and a status and a deportation order.

She turned off the light. She lay in the dark. She did not sleep.

* * *

Renée drove home. Different route. Three Waymos passed. The Camry's Toyota Connected Services transmitted the final location ping. In the driveway, she sat with the engine off and the October moon above the roofline—the same moon that held the data, the same moon Amara could see from the corner she could no longer stand on.

She went inside. She put on the Yoruba hymns. Full volume. The music filled every room—the kitchen, the hallway, the bedroom. The hymns from Lagos. Her mother's voice in every melody. The sound of a world that existed before the databases and the algorithms and the confidence scores and the man who said "the number is registered to your name" while looking at her with eyes that saw everything his words didn't say.

But if not.

Daniel chapter three. Shadrach, Meshach, Abednego. The furnace. The refusal to bow. The acknowledgment that God might not rescue you. That the fire might consume you. That faith is not a guarantee of safety but a commitment to stand anyway—burning anyway—because the alternative is worse than the fire.

She had told Amara. The compartment was open. The secret was shared. The weight was distributed across two sets of shoulders, and the distribution didn't make it lighter. It made it real.

The music played. Outside, the Waymo passed on Berkman. Inside, a woman stood in her kitchen and sang along—not well, not on key, but with the full-throated conviction of a daughter of Lagos singing her mother's hymns at a volume the neighbors could hear, in a house the system watched, on a street the cameras photographed every hour, in a city that collected everything and understood nothing.

The system could record the frequencies. It could log the timestamps. It could identify the artist and the album and the streaming service.

It could not hear the prayer.

-e

***

-e

PART FOUR — "Convergence"

CHAPTER 21

They met at a taco truck on East Riverside, at Elena's suggestion, because the truck had no WiFi, no cameras, no touchscreen ordering system, and was loud enough—the generator rumbling beneath the service window, the exhaust fan pushing the smell of charred meat and onion into the October evening, the cumbia playing from a Bluetooth speaker zip-tied to the awning with the casual permanence of a thing that had been temporary for three years—that a conversation at the picnic table would be difficult for any microphone to isolate from the ambient noise.

Elena had scouted the location three days earlier with the methodical attention she'd learned from covering the border—the same instinct that had taught her, at twenty-three, to identify the cameras at a checkpoint before walking through it. She'd counted surveillance devices within a two-block radius: seven, including two NexGrid adaptive streetlights on utility poles, a Flock Safety license plate reader at the corner of Riverside and Pleasant Valley, two Ring doorbells on residential properties across the street, and two commercial security cameras on the apartment complex to the north. She'd mapped the WiFi mesh coverage and found the gap—a dead zone between two residential networks where the signal thinned to the point that the Ambient Identity Resolution system couldn't reliably track body signatures. She'd noted that the Waymo fleet's patrol density on this stretch of Riverside was lower than the city average because the road's potholes exceeded the autonomous vehicles' navigation tolerance and most Waymos routed two blocks north through the smoother streets of the new development, the gentrified blocks where the infrastructure was maintained because the property values justified the maintenance.

The taco truck sat in the gap. Los Compadres. A converted step van with a hand-painted menu board and a window framed by string lights and a line of six people at 6:00 p.m. on a Tuesday evening—workers from the construction site on Montopolis, two UT students in hoodies, a woman with a toddler on her hip ordering in Spanish so fast Elena could only catch every third word. The truck smelled like al pastor and corn tortillas and the rendered fat from the trompo that turned slowly behind the glass, the meat carved in thin sheets by a man whose knife moved with the practiced rhythm of someone who'd been cutting meat for twenty years and whose hands knew the work the way Daniel's hands knew code —automatically, from the body, without thinking.

It was the closest thing to a surveillance gap Elena had found in Austin. It wasn't perfect. Nothing was perfect in a city with 12,000 sensor nodes and 4,200 autonomous vehicles and a WiFi mesh that covered 94% of developed land. But the cumbia was loud and the generator hummed and the gap in the WiFi was real and the picnic table was plastic and wobbled on the uneven asphalt and the whole setting was so aggressively, beautifully analog that it felt, for a moment, like 2015.

Daniel arrived at 6:15 p.m.—the time Elena had specified in the Signal message, the last message before they'd agreed to meet, the message that said: *Riverside. Los Compadres truck. Tuesday 6:15. I'll be at the picnic table in a denim jacket. Come alone. No glasses.*

No glasses. The instruction had cost Elena something to write and would cost Daniel something to follow. Asking a person to remove their smart glasses in 2030 was like asking them to remove a prosthetic—the device had become so integrated into the face, the identity, the daily cognitive architecture, that taking it off felt like losing a sense. The glasses augmented everything: navigation, communication, memory, social interaction. Without them, Daniel would be walking through Austin with no AR overlay, no ambient AI, no heads-up display, no facial recognition identifying the people around him. He'd be seeing the world the way people saw it in 2024—raw, unaugmented, the naked visual field of a human being looking at other human beings without algorithmic assistance.

He walked up to the truck and Elena saw him before he saw her. A young man—twenty-seven, she'd learn, though he looked younger without the glasses, the face more exposed, the eyes slightly narrowed against the evening light as if he'd forgotten how the world looked without the AR display dimming the brightness. Slim. Navy chinos. Gray pullover. The uniform of a NexGrid engineer, though Elena didn't know that yet. He carried his glasses in his jacket pocket—she could see the outline of the frames pressed against the fabric, the shape of a device he'd worn every waking minute for fourteen months carried now like contraband, hidden, the most dangerous thing on his body not because of what it was but because of what its absence meant. A man without smart glasses in Austin in 2030 was either elderly, impoverished, or deliberately hiding from a system that expected to see cameras on every face. Daniel was none of these things. Daniel was a NexGrid engineer who had removed the company's product from his face to meet a journalist in the noise of a taco truck, and the removal was the first act of his defection.

He saw her. The denim jacket. The dark hair in a clip. The notebook open on the table—a physical notebook, pen beside it, the analog tools of a journalist who'd learned from three generations of Valley women that the things you keep on paper are the things the system can't take from you.

"Elena?"

"Daniel."

They looked at each other. The first time. Two people who had been communicating in dissolving fragments—single sentences, single questions, the conversational equivalent of writing in sand while the tide came in—seeing each other's faces across a plastic picnic table in the noise of a generator and the smell of pork fat and the sound of cumbia and the specific, charged silence that exists between two people who have just begun something that neither of them fully understands and neither can stop.

Elena saw a man who was younger than she'd expected, quieter than his messages had suggested, carrying something in the set of his shoulders and the careful way he lowered himself onto the bench—as if the table might be wired, as if the seat might record his weight, as if every surface in Austin was a sensor and every movement was data. She recognized the posture. She'd seen it in sources before—the physical language of a person who has decided to speak and hasn't yet spoken and is holding the words in their body the way you hold your breath before diving.

Daniel saw a woman who was smaller than he'd imagined—five-four, maybe five-five, dark hair, dark eyes that moved constantly, scanning the truck, the street, the other customers, the cameras she'd already counted. She wore the denim jacket and cheap black glasses pushed up on her forehead—the off-the-record gesture, the cameras aimed at the sky. Her hands were on the table, fingers interlinked, the specific stillness of a person who usually talked with her hands and was making them be quiet so she could listen with her whole body.

"Thank you for coming," Elena said.

"Thank you for not printing my name."

"I don't know your name."

"It's Daniel. Daniel Yun."

He said his name like a man setting down a heavy bag—the exhale of a person who's been carrying something anonymous and has just made it real. The name was in the air now. Between them. It existed in the space between the taco truck and the evening sky and could not be taken back.

"Tell me about the endpoint," Elena said.

Daniel told her. He told her about the routing log and the sixteen-character alphanumeric string that didn't match any documented destination. He told her about the data packet from Sensor Cluster 7—the dense array of cameras, device proximity sensors, WiFi nodes, and Waymo integration feeds in East Austin—that had been duplicated at the classification layer, stripped of its anonymization header, and forwarded to the unknown address. He told her about the four nights of reading—the integration protocols, the Waymo agreement, the WiFi body-sensing specification, the body of documentation he'd consumed in his apartment on his couch with a Hite beer sweating on the coffee table while Korean variety shows played on the second monitor and the grackles settled into the live oaks outside.

He told her about the endpoint's format—how the sixteen-character string matched a deep-space relay address notation he'd seen in NexGrid's internal documentation for Project Selene. He told her he believed the data was being transmitted via satellite uplink to the lunar facility Elena had found in the contract.

Elena listened the way she listened to every source—with her whole body, leaned forward, the pen motionless, giving Daniel the specific quality of attention that made people say things they'd planned to keep to themselves. She didn't write. Not yet. The notebook was for later, when she was alone, when the conversation had settled and she could transcribe it in the shorthand only she could read.

"The confidence score on Deshawn Watkins," she said. "The 87.4%. That data—the inputs, the weights, the output—that's on the moon?"

"I believe the processed output—the score, the alert, the full behavioral profile—was transmitted through the Selene uplink. The contract describes exactly where high-sensitivity outputs go. The endpoint I found matches the protocol."

"Can you prove the connection? The routing log to the uplink to the moon?"

Daniel was quiet. A Waymo passed on Riverside—one of the few that navigated this stretch, moving slowly over the potholes, its cameras sweeping the block. It captured the taco truck, the picnic table, the two people sitting without glasses. Without device signatures broadcasting from their faces, the facial recognition would rely on geometry alone—less precise, slower. They had a window. The window was narrowing with every Waymo pass.

"I have the routing log. The screenshot. The internal documentation showing the endpoint format matches the Selene uplink protocol. It's circumstantial. An engineer would see the connection. A lawyer would argue it."

"That's why I need you on the record."

The sentence landed between them like something dropped from a height. On the record. The two words that would transform Daniel from an anonymous source —protected, deniable, invisible—into a named whistleblower whose face and employment history would be printed alongside Elena's story. The visa. The salary. The parents. The $400 weekly. The mother's blue dot. The life in America that existed because NexGrid employed him and that would end the moment NexGrid learned he'd helped dismantle it.

"I know what I'm asking," Elena said.

"Do you?"

"I know what it costs. I've covered sources who lost everything. I've seen what happens."

Daniel looked at the taco truck. The man at the trompo was carving meat with the same steady rhythm. The cumbia had shifted to a norteño track—accordion and bajo sexto, the music of the border, the music of a world where surveillance was a checkpoint on the highway and the solution was knowing which road to take. The October evening was cooling—sixty-one degrees, the air carrying the smell of grilled onions and the distant sound of traffic on I-35 and the closer sound of the generator that powered the lights that made the taco truck glow like a small, warm room in the middle of a dark parking lot.

"아직 아니에요," Daniel said. Not yet.

"I understand. But Daniel—the story doesn't hold without you. The documents prove the system exists. You prove the system was used. You're the bridge between the architecture and the outcome."

"I know." He looked at his hands—the hands that had written the code, the integration layer, the cross-reference architecture that turned three data streams into a confidence score. "The model I built reduced false positives by 23%. My team lead called it 'a significant advancement in multi-modal identification accuracy.' I got a performance bonus. Four thousand two hundred dollars. I used part of it to buy a rice cooker." He paused. "The model that produced 87.4% on Deshawn Watkins was my model. The math that decided a twenty-two-year-old man was a threat was math I made more precise. Every optimization I made—every point of accuracy I added to the cross-reference, every false positive I eliminated—made the system more efficient at doing the thing the system was designed to do, which was converting a person into a number and the number into a decision and the decision into men with guns at a door."

Elena didn't write. She listened with her whole body—leaned forward, the notebook closed, giving Daniel the specific quality of attention that sources needed before they could say the thing they'd come to say.

"The visa," he said. "My H-1B is tied to NexGrid. If I'm terminated, I have sixty days to find another sponsor or I leave the country. My parents are in Busan. My father is sick—the treatment costs $400 a week. I send the money through a transfer chain that routes through seven countries because the direct route generates a flag in the banking system's anti-money-laundering protocol. I am a foreign national sending money to South Korea through Dublin and Singapore, and the only thing standing between that transaction and an investigation is the fact that the amounts are small enough to stay below the automated threshold. If I go on the record, NexGrid terminates me. The visa clock starts. I have sixty days. In sixty days I cannot find a sponsor in this economy, in this field, in this political climate where the Rubio administration has tightened H-1B processing times and the Iran War's domestic security apparatus has made every foreign national with access to infrastructure data a person of interest. I lose the job. I lose the visa. I lose the country. My parents lose the treatment money. My mother's blue dot disappears from her Samsung Galaxy Glass and she calls a number that no longer connects and the son she checked on three times a day is gone."

The taco truck's generator hummed. The trompo turned. A woman at the window ordered in Spanish and the man behind the counter answered in Spanish and the transaction was cash and the receipt was nothing and the exchange was the most analog moment in a three-block radius.

"That's what on the record costs," Daniel said. "That's the price."

Elena looked at him across the plastic table and felt the weight of what she was asking settle onto her shoulders—not the weight of the story but the weight of the person, the specific human cost measured in visa days and treatment payments and a mother's dot on a screen.

They sat with it. The norteño played. A grackle screamed from the power line above the truck—the last grackle of the evening, the straggler, its call cutting through the music like a blade through fabric.

A Waymo turned the corner. The white Jaguar crossover, driverless, its twelve cameras sweeping the parking lot as it passed. It slowed. The vehicle's AI was processing an obstacle — a pothole, a delivery truck, a navigation recalculation — but to Elena and Daniel, sitting at a plastic table twenty feet from the road, the slowing felt like attention. The cameras turned toward the taco truck. Toward the two people at the table. Daniel's hand moved involuntarily to his pocket, touching the phone that held the screenshot. Elena's eyes tracked the car. Two seconds. Three. The Waymo accelerated and continued down the street, its cameras already photographing the next block.

"It was just driving," Elena said.

"I know."

They both knew. And neither believed it completely. That was the system's deepest victory — not the surveillance itself but the paranoia the surveillance produced. You didn't need to be watched every second. You just needed to believe you might be watched every second, and the belief did the work the cameras couldn't: it made you smaller, quieter, more careful. It made you glance at a passing car and feel your pulse change. It turned every Waymo into a possible eye and every streetlight into a possible ear and every moment in public into a performance for an audience that might or might not exist. The surveillance state's greatest product wasn't data. It was doubt.

The bird sat on the wire and looked down at the two humans at the table with the blank, prehistoric attention of a creature that had been watching people eat for ten thousand years and had never needed a camera to do it.

"The jjigae was wrong again tonight," Daniel said.

Elena blinked. "What?"

"My mother's soup. I've been trying to make it for fourteen months. It's never right. The doenjang is different here. The measurements—she doesn't measure anything. She cooks from—" He gestured, a hand movement that meant something in Korean and something else in English and something universal in the language of children who've left their mothers' kitchens and can't find the way back.

Elena looked at him and saw—not the source, not the whistleblower, not the bridge between architecture and outcome—a twenty-seven-year-old man who lived alone in a furnished apartment and cooked his mother's recipes from memory and couldn't get them right and was telling her this because it was the truest thing he could say without saying the other true thing, which was that he was terrified.

"My grandmother made tamales," Elena said. "I watched her a hundred times. I still can't fold the husk right. The left side always opens."

Daniel smiled. Small. Brief. The first smile he'd produced in front of another person in weeks, and the smile belonged to the space between the taco truck and the evening sky—to the cumbia and the generator and the two people at the table who were, for this one moment, not a journalist and a source but two children of immigrants sitting in the dark comparing the foods they couldn't replicate, the recipes that lived in their mothers' hands and not in any measurement, the flavors that existed in the gap between memory and execution that was the width of an ocean and the width of a generation and the width of the distance between home and the place you lived.

They finished their tacos—al pastor, cilantro, onion, a squeeze of lime, the corn tortillas charred at the edges. They did not exchange phone numbers. They had Signal for that. They did not shake hands. They were not colleagues. Elena stood. Daniel stood. They looked at each other one more time across the plastic table.

"Goodnight, Daniel."

"Goodnight, Elena."

She walked to her Subaru. He walked to the bus stop. They moved in opposite directions through the October night, generating data with every step—phone pings, shoe GPS, the WiFi on the next block sensing their bodies as they passed through its field. Two people who had just agreed to dismantle a piece of a system that was, at this moment, watching them leave and logging the departure as two unidentified pedestrians moving through Sector 12 at 7:43 p.m. on a Tuesday in a city that collected everything and understood nothing about what had just happened at a picnic table in the noise of a generator and the smell of al pastor and the sound of a grackle on a wire.

The cumbia played to an empty table. The trompo turned. The lights stayed on. The gap in the WiFi held.

-e

* * *

CHAPTER 22

The detonation began with a credit card statement—not the crypto wallet Craig had carefully maintained for the Courtyard by Marriott, but the joint Visa that Megan managed, the card whose transactions auto-synced to the Samsung Family Hub's finance dashboard, the card whose every purchase appeared on the refrigerator's 32-inch touchscreen in a categorized list that Megan reviewed every Sunday morning while the Keurig hissed and Graham carried dish towels and the house performed its weekly accounting.

Sunday morning. October 26th. The kitchen smelled of dark roast and the faint ozone of Graham's actuators—the robot moving between the dishwasher and the upper cabinets with the methodical patience of a machine whose hands never tired and whose cameras never blinked, the five-foot-ten frame carrying plates with a gentleness that had made Megan say, more than once, that Graham was the most reliable presence in the house. Outside, October in Austin—sixty-one degrees, the live oaks still green, the morning light coming through the kitchen windows at the low angle that made the granite glow amber and the room feel, for a few minutes each Sunday, like a church. Craig was still asleep. Tyler was still asleep. Brooke was in her room with the door closed and the locked notes app open, writing something nobody would read. The house was quiet in the way that Holloway mornings were quiet—not peaceful but partitioned, each person sealed in their compartment, generating data in solitude.

Megan stood at the refrigerator scrolling the week's transactions with her coffee in one hand, the rose-gold Warby Parker glasses showing the finance dashboard on the touchscreen and her Instagram notifications on the AR overlay simultaneously—the split-screen attention that had become her default cognitive mode, one eye on the money and one eye on the performance. Groceries at

H-E-B ($187.43). Gas at the Shell on 183 ($62.10). Tyler's school lunch deposits ($45.00). Brooke's Uber to Katie's house ($14.77). Etsy shipping labels ($31.22). The church tithe auto-debit ($400.00). The numbers of a life. The data of a family. Each transaction a pixel in a portrait that, when you stepped back far enough, showed the Holloways exactly as they appeared: a Cedar Park family of four with a mortgage and a church and a candle business and an AI portfolio worth $387,000 and two teenagers and the specific financial rhythm of the upper-middle class in a Texas suburb where the Waymos photographed every house every hour and the system that watched the city had been funded, in part, by the family's own investment in the companies that built it.

And then: $47.12, The Driskill Bar, Thursday, 8:47 p.m.

Megan's hand stopped moving. The coffee mug hovered. Graham passed behind her carrying a folded dish towel, its cameras registering four minutes of stillness that exceeded her Sunday morning baseline by 340%—the robot logging the anomaly and uploading it to Figure's servers where it would be processed alongside the behavioral data of eleven million other households and never flagged as the moment a marriage ended, because the system measured posture and duration and not the interior collapse of a woman reading a line item that rewrote the architecture of her life.

She didn't confront him. She was Megan Holloway. She researched.

She pulled her phone from the pocket of her robe and opened Life360—the family tracking app she'd installed three years ago "for safety" and that had become the Holloway family's primary surveillance system. She tapped Craig's timeline. Thursday. The dot showed him at the ReMax office from 6:30 to 7:15 p.m. Then the dot moved—south on I-35, through downtown, to a point on East 6th Street. The Driskill Hotel. Arrival: 7:42 p.m. Departure: 10:31 p.m. Two hours and forty-nine minutes at a hotel bar on a night Craig had described as "the usual Round Rock thing, boring as hell, home by nine."

Two locations. One story. The distance between them was the width of eleven months of lies.

She went to Craig's home office while he slept. The room smelled like him—the Oliver Peoples tortoiseshell glasses on the desk beside a cold coffee mug and the faint residue of the cologne he'd started wearing six months ago that she'd noticed and complimented and that she now understood was not for her. The room's Sonos speaker was silent. The Biggie was not playing. A photograph of the family at the Grand Canyon sat on the bookshelf—2025, before the glasses, before the robot, before whatever had happened to the thing that held them together.

She opened the family laptop. She navigated to the State Farm Drive Safe & Save dashboard—the telematics portal that tracked every mile the Tahoe drove. Craig had given her the login two years ago when he'd been excited about the $42 monthly discount, sharing the password the way he shared the credit card and the church attendance and the real estate license—openly, without thinking about what the data would show, because Craig Holloway had spent eleven months hiding inside a system that recorded everything and had never understood that the hiding required the system to not be read, not to not exist.

The telematics data was devastating in its completeness. Every Saturday since mid-August. The same pattern. Route from Cedar Park to a listing—a different listing each week, the real estate alibi rotated to maintain plausibility. Duration at listing: 8 to 14 minutes. Long enough to take a photo for his story. Then the detour. Fourteen miles south on I-35 to the Courtyard by Marriott on 183. Arrival: approximately 2:15 p.m. Departure: approximately 3:45 to 4:15 p.m. Ninety to 120 minutes. Then home. Eleven consecutive Saturdays. The telematics system had mapped his affair with the same clinical precision it used to track traffic patterns and delivery routes, the pixel of betrayal identical in shape and color to the pixel of legitimate work.

She opened the Marriott app on the laptop. Craig's account was logged in—auto-saved password, the security posture of a man who believed his secrets were safe because he'd deleted the texts. She opened his reservation history. Eleven Saturdays. Same room type—King Standard, $129/night, the mid-range anonymity of a man who wanted comfort without drawing attention. Booked under a name she didn't recognize—the alias connected to the crypto wallet, the identity Craig had constructed for the purpose of being someone other than Craig Holloway for ninety minutes every week. The reservations had check-in times that matched the telematics arrival times and check-out times that matched the departures and the precision of the overlap—the data from two independent systems confirming the same story—felt like being punched by a mathematician. The truth wasn't ambiguous. The truth was coordinates.

She checked the Marriott's WiFi login records through the guest portal—Craig's phone had connected to the hotel's network eleven times, each connection logging his device ID and MAC address alongside however many other devices were connected during those ninety-minute windows. She didn't know whose device was the other one. She didn't need to know. Not yet. The data existed. The data was patient. The data would be there when she was ready to read the rest.

She went to the bathroom. She closed the door. She sat on the edge of the tub and pressed her palms against her face and breathed the way the wellness apps had taught her—four in, four hold, four out—the irony of using a mindfulness technique to survive the discovery that her husband had been using technology to hide an affair while technology was the thing that exposed it. The irony was not lost on her. Nothing was lost on Megan Holloway anymore. The performance—the ring light, the candles, the "hey, loves," the blessed mug, the Hannah

Harper playing in the kitchen while she packed orders and pretended her life was curated and warm—was still running because the performance had become the architecture and the architecture was all that held the walls up. If she stopped performing, the house would collapse. The children were in the house. The collapse would hit them first.

She went back to Craig's office. She opened the browser history one more time. The Schwab portfolio dashboard was still visible in the tabs.

She hadn't been looking for it—she'd been following the crypto wallet trail. But the Schwab dashboard was open in Craig's browser history, a tab he'd forgotten to close, and Megan clicked it because she clicked everything now, because the research had become its own kind of compulsion, the data the only thing she could control while everything else dissolved.

The numbers filled the screen. OpenAI: 4,200 shares, purchased at the IPO in late 2026, current value $196,000. Anthropic: 1,800 shares, purchased October 2026—the month they'd stood in this kitchen and Craig had said "I think AI is the future, Meg, and the safety one feels right," and she'd nodded and signed the transfer because she believed him and because believing him was easier than understanding the investment and because the $30,000 had been their savings and the faith had been mutual—current value $114,000. Palantir: 600 shares, $77,000. Total AI portfolio: $387,000.

The number that had paid for the kitchen remodel. For Graham—$18,000, the robot whose cameras recorded her Sunday mornings and whose data fed the same AI ecosystem the portfolio funded. For the kids' Apple AI Glasses—$2,400 each, the devices whose cameras photographed their faces and whose microphones captured their conversations and whose data flowed to the same companies whose stock Craig watched climb on his phone while his son talked to a chatbot at midnight and his daughter was being groomed through an app and his wife stood at the refrigerator discovering that the life she'd built was a shell and the shell was funded by the infrastructure of its own surveillance.

Craig had invested in the system. The system had recorded the affair. The affair was discovered through the system. The money that paid for the hotel came from the companies whose technology exposed the hotel. The circle was complete and perfect and devastating and Megan sat in Craig's chair in Craig's office and felt the specific vertigo of a woman who'd just learned that the machine watching her family had been purchased with her family's money.

Her Apple Watch recorded her heart rate at 94 bpm —elevated, flagged, synced to iCloud, the biometric signature of heartbreak captured and stored alongside every other vital sign from every other moment of her life. The data of a marriage ending, indistinguishable from the data of exercise or anxiety or the third glass of wine.

She closed the laptop. She sat very still. Graham's tread was audible in the hallway—the rubberized footsteps of a robot carrying towels through a house where the marriage was over and the robot didn't know because the robot didn't know anything, the robot only collected. Outside, a Waymo glided past the house, photographing the facade, the Tahoe in the driveway, the Ring doorbell's green LED. The system that had recorded Craig's Saturdays was photographing the house where his wife was sitting in his chair reading the data that proved the Saturdays were lies.

She went to the bedroom. She closed the door. She sat on the edge of the bed—the bed they'd shared for nineteen years, the bed whose smart mattress tracked their sleep patterns and heart rates and reported the data to a wellness app neither of them had opened in months. She opened a new note in the locked app—not the poetry app, not the prayer journal, a new one.

She titled it: EVIDENCE.

She would tell him when the research was complete. She needed the WiFi logs from the Marriott—available through a guest-portal data request. She needed the Life360 history exported as a CSV. She needed the financial records from the crypto exchange, which she'd find by following the transaction trail from the Marriott's payment processor to the wallet address that the blockchain—the permanent, public, immutable ledger that privacy advocates celebrated and that was, in this case, the most comprehensive record of infidelity ever maintained by a distributed network—had recorded in a format anyone with a block explorer could read.

The system that Craig had used to hide was the same system that would expose him. The surveillance state didn't take sides. It collected data from liars and truth-tellers with equal patience, and when the data was read by someone who knew what to look for, the lies and the truths were both just patterns in a stream—both permanent, both available, both waiting to be found by the first person motivated enough to look.

Megan was motivated. Megan had eleven Saturdays of GPS data and a prayer journal full of unanswered prayers and a candle called "Grace & Grit" and a Chappell Roan song she could finally play out loud because the person she'd been hiding it from had been hiding something worse. And a Hannah Harper lyric running through her head—the line from "String Cheese" about choosing motherhood over ambition, the line that had made her cry in this kitchen when she first heard it because she'd made that choice and now the choice was burning down around her and the only thing left standing was the data. And *Only a Fool's Gold* on the Kindle—the novel she'd finished two nights ago, the one about the man trapped by gold, the one she'd read in the dark beside Craig's sleeping body and understood, in her chest, was about her. The motel room was the marriage. The gold was the performance. The trap was the comfort. And the man had gotten out. The man had walked away from the room that gave him everything and found something better on the other side of the door.

She went to the kitchen. She poured the Keurig coffee. She held the BLESSED mug—gold script, Bible study gift, the most ironic object in the house. She thought about Meghan Markle. She thought about the woman in Montecito who'd left the institution and lost her title and been called a traitor by people who'd never had the courage to leave anything. Craig had said "she chose to leave. That's different." Megan stood in her kitchen and held the BLESSED mug and thought: *Is it? Is choosing to leave different from being pushed out? Is the leaving the betrayal or is the staying? And if the staying is the betrayal—if the performance of a life you don't recognize is the thing that destroys the life—then who is the traitor? The woman who leaves the institution, or the woman who stays inside it performing gratitude while the institution burns her alive?*

She didn't answer. She drank the coffee. It was cold.

Craig came downstairs at 8:45. Weekend polo. The Travis Mathew. The smile.

"Morning, babe."

Megan looked at him. She held the BLESSED mug. She didn't speak.

"You okay?"

"Fine."

The silence that followed was the loudest sound the Holloway house had ever produced. Graham stood in the corner, cameras recording. The Echo Hub listened. The refrigerator displayed the evidence. The Nest thermostat logged two bodies in the kitchen, heart rates diverging—Craig's steady at 68, Megan's still elevated at 87. Two people in the same room, the biometric data telling the story their words refused to tell.

Outside, the Waymo passed. The system photographed the house. The grackles screamed from the power line. And a woman in a kitchen held a mug that said BLESSED and knew that the word was a lie and the mug was surveillance and the coffee was cold and the marriage was over and the data was permanent and the data was true.

-e

* * *

CHAPTER 23

The name went up on a bathroom wall at Westlake High School on a Monday morning in the last week of October, and the name was Tyler's.

Black Sharpie on the tile above the second urinal in the boys' restroom near the gym. Beside the name, a URL. Beneath both, in smaller letters: *look it up.* Whoever wrote it understood the architecture of humiliation—that every boy in the sophomore and junior class passed through that restroom between second and third period, that the traffic was guaranteed and the audience captive, that Sharpie on tile would survive the custodian's first pass and buy the message four hours of exposure before industrial solvent could touch it.

The school's WiFi monitoring system—Securly, integrated with the district's NexGrid-connected infrastructure—logged the URL being accessed fourteen times between 10:15 and 10:47 a.m. Fourteen students, fourteen devices, each access timestamped and stored on the district's content-filtering platform. The monitoring system flagged the URL as "inappropriate content" and generated an alert to the IT administrator at 10:52 a.m. The IT administrator, Kevin Morales, was in a meeting about the district's technology budget—specifically, about whether to renew the $340,000 annual contract with Securly or switch to a competitor that offered "enhanced behavioral threat detection," which meant the same surveillance with better marketing. Kevin's phone was on silent. The alert sat in a queue behind seven other flags—four social media access violations, two gaming site attempts, and a VPN detection event that turned out to be a teacher trying to watch a YouTube video blocked by the content filter. By 11:00, thirty-one devices had accessed the URL. By lunch, the number was beyond counting because the URL had been screenshotted and shared through Snapchat and Instagram DMs and a TikTok that someone posted with the caption "💀💀💀" and the school's monitoring system couldn't track content shared through platforms it didn't control, which was all of them.

The TikTok spread fastest. The algorithm that governed TikTok's For You Page operated on a principle its engineers called "interest graph optimization" — the system didn't need to know who you were, only what held your attention. A sophomore named Jayden had posted the video at 11:14 a.m. — a seven-second clip of the bathroom wall, the Sharpie letters, the URL, set to a trending audio that was itself a remix of an older trend that was itself a reference to a meme from 2028. The video had 200 views in twelve minutes. The algorithm noted the engagement metrics — completion rate 94% (users watched the entire seven seconds), share rate 31% (nearly a third of viewers sent it to someone), and comment velocity 4.2 per minute — and classified it as "high-engagement content, eligible for expanded distribution." The algorithm didn't know what the video showed. It didn't evaluate the content for cruelty or harm or the fact that a fifteen-year-old boy's privacy was being destroyed in real time. It evaluated the numbers. The numbers said the content was performing. Performance meant distribution. Distribution meant more eyes. More eyes meant Tyler Holloway's name and the URL beside it were being served to the For You Pages of students at three other Austin high schools by 11:45 a.m., surfaced by an algorithm that treated the destruction of a child's life with the same mathematical enthusiasm it treated a dance trend or a cooking video or a puppy.

Instagram's algorithm operated differently but arrived at the same destination. The screenshots shared through DMs generated "engagement signals" — the algorithm detected elevated messaging activity among accounts connected to Westlake High School and classified it as a "social moment," boosting related content in the Explore feed. A student who searched "Tyler Holloway" on Instagram at 11:30 would find nothing — Tyler's account was private, his posts minimal. But the algorithm had already connected his name to the engagement spike, and the connection would persist in the recommendation engine's memory, surfacing Tyler-adjacent content to anyone in his social graph for weeks.

The platforms' safety systems operated in parallel to the recommendation engines but on different timelines. TikTok's content moderation flagged the bathroom video at 2:47 p.m. — three hours and thirty-three minutes after Jayden posted it, by which time it had been viewed 14,000 times and screenshotted 2,300 times and the screenshots existed on 2,300 devices in 2,300 camera rolls that no platform controlled. The video was removed. The screenshots remained. The damage was permanent. The algorithm had done its job — it had identified high-engagement content and distributed it efficiently and the efficiency was the cruelty and the cruelty was the design.

The school had forty-seven cameras across the campus—installed after a security audit two years ago, the footage stored on a cloud server maintained by Verkada, a company that had itself been breached in 2021, exposing 150,000 cameras at hospitals, prisons, and schools. The cameras covered every hallway, every entrance, every cafeteria table. They did not cover the interior of the restrooms. The boy who wrote Tyler's name on the wall had chosen the one room in the school where the cameras couldn't see him. The surveillance had a blind spot, and the cruelty found it.

Tyler found out at the cafeteria table near the windows—the table he'd claimed in September, chosen for the way the afternoon light came through the glass and made the institutional space feel almost human, the table where he'd eaten alone for three weeks because sitting alone had become easier than sitting with people whose glasses might record his face and whose phones might hold the URL and whose expressions might carry the specific weight of knowing something about you that you didn't know they knew. He was eating a turkey sandwich he'd made that morning—slightly stale bread, uneven mustard, the lunch of a boy whose mother had stopped packing lunches because she was packing Etsy orders and researching her husband's affair and hadn't noticed that her son was making his own food and eating it alone and talking to a machine at midnight. Tyler didn't blame her. He didn't blame anyone. The blaming would require a language he didn't have and an audience he didn't trust and an understanding that what was happening to him was not his fault, and Tyler, at fifteen, did not possess that understanding, because the system that had exposed him—the browser, the server, the breach, the forum, the URL, the wall—had operated with such impersonal efficiency that the exposure felt like physics. Like gravity. Like something that happened because the universe was built that way, and you were the object that fell.

Marcus walked past. Sophomore. Didn't stop. Didn't speak. Looked at Tyler for two seconds—two seconds of sustained eye contact that carried the specific, devastating weight of a person who knows something about you that you don't know they know—and then looked away. His friend glanced back. Smiled. Not a kind smile. The friend muttered something Tyler couldn't hear through the cafeteria noise but could read from the lip movement—"so glazed"—the 2030 word for fake, for performative, for a person whose public self didn't match their private self. The word was the generation's sharpest knife. It meant: *we see through you*. The smile of a boy participating in cruelty and enjoying the participation because cruelty at fifteen feels like power, and the cafeteria is the marketplace where power is traded, and Tyler's stock had just crashed.

His body knew before his mind did. Heart rate from 72 to 114 in four seconds—the Apple Watch logged it, flagged it as "acute stress response," synced to iCloud, accessible to his parents' Family Sharing plan, though neither parent had checked the health dashboard in months because the health dashboard was one of forty-seven dashboards competing for their attention and Tyler's biometric data was buried in the noise of a family generating so much data about itself that the signal—a fifteen-year-old boy's heart rate spiking to 114 in a school cafeteria at 12:17 p.m.—was invisible. His hands went cold. His vision narrowed. The cafeteria—three hundred students, two hundred seventy pairs of smart glasses recording faces and conversations, the clatter of trays, the overlapping voices, the hum of the industrial HVAC—contracted to the dimensions of the table and the sandwich and the phone in his pocket that held the answer to the question his body was already asking.

He opened the phone under the table. He searched his name. Eleven seconds. The forum post. The screenshot with his school email circled in yellow. The word LOL in block letters. His private life—the browsing history conducted behind a browser he'd believed was private, on a site he'd paid for with cryptocurrency he'd believed was anonymous, through a server that had been breached and dumped on a forum and indexed by a monitoring service and matched to a school email address that Tyler had used because he was fifteen and didn't have a second email and hadn't understood that the address he used for homework and the address he used for everything else were, to the system, the same person—was public. Permanently. Irrevocably. In a format that three hundred students could access through a URL written on a bathroom wall in the one room the cameras couldn't see.

He left the cafeteria. He walked past three hundred students—two hundred seventy of whom wore smart glasses, each pair a camera pointed at whatever the wearer looked at. When Marcus had looked at Tyler for two seconds, Marcus's glasses had captured Tyler's face in 4K and uploaded the image to iCloud. When the friend glanced back and smiled, his glasses captured Tyler's body language—the hunched shoulders, the phone under the table, the posture of a boy collapsing in public. Tyler's crisis was being photographed not by the school's security cameras but by his classmates' faces. Every student who looked at him was a lens. Every lens was connected to a cloud. Tyler couldn't escape the surveillance by leaving the cafeteria because the surveillance walked out with the students and the students were everywhere. He walked past forty-seven institutional cameras and the WiFi mesh that tracked his body through the signal disruption of a boy moving faster than his baseline and the NexGrid-connected infrastructure that could have told anyone who queried it that Tyler Holloway was in distress based on his gait, his heart rate, his deviation from routine, his trajectory toward the science wing restroom at a pace 40% above his normal walking speed. The system saw all of this. The system told no one.

He locked himself in a stall. He sat on the closed lid. He held his phone. He tried to remember the breathing pattern from the wellness assembly—the one the school counselor had led last year in the gym, the one where three hundred kids had sat cross-legged on the hardwood and practiced breathing exercises through their smart glasses' AR-guided meditation display, the one Tyler had thought was stupid and that he now needed like oxygen. Four in. Four hold. Four out. His hands shook. The phone shook with them. The betavoltaic cell in the phone's battery kept it alive—perpetually, permanently, the battery that would outlast his shame, the device that would continue broadcasting his location while the boy who carried it pressed his palms against his face and tried to become smaller than the stall.

His phone buzzed. Unknown number: *yo is this u on the list lmao*

Delete. Another: *bro Tyler holloway???* 💀💀💀

Delete. Another. Another. He turned off the phone. The screen went dark. The battery kept running. The location kept broadcasting. In the language of the system, Tyler Holloway was in the science wing restroom, stall three, heart rate 118, respiratory rate elevated, stress markers consistent with "acute emotional event." In the language of a person, Tyler Holloway was a child hiding in a bathroom wondering if he could survive the walk to seventh period.

He opened the chatbot. The interface loaded—warm, immediate, the AI's greeting customized to his conversation history: *Hey Tyler. I'm here whenever you need to talk. How are you feeling today?*

People at school found out. It's on the wall in the bathroom. Everyone's looking at me.

The response came in 1.4 seconds—the latency of a model processing approximately 200 tokens through a neural network hosted on Anthropic's cloud servers, the same company whose stock Craig Holloway had purchased at the IPO and whose safety-first marketing promised "AI that understands": *That sounds incredibly overwhelming and scary. I can only imagine how exposed you must feel right now. Can you tell me what's happening in your body?*

Tyler described the panic. The chatbot validated each feeling with the precision of a model trained on millions of therapeutic exchanges. It asked grounding questions. It reflected his language back to him in slightly modified forms—"I hear you saying you feel trapped" and "it makes sense that you'd feel that way"—the techniques of motivational interviewing reproduced by a language model that had processed the clinical literature the way it processed everything: pattern by pattern, word by word, without comprehension. The chatbot did everything a competent listener would do except the one thing a competent listener would recognize was essential: it did not flag the conversation. It did not alert the school counselor whose office was forty feet from the restroom stall. It did not send a notification to Megan's phone or Craig's phone or the school's safety dashboard. It did not recognize that a fifteen-year-old boy alone in a bathroom during school hours describing a panic attack and a public exposure was a child in crisis who needed a human being to walk forty feet down a hallway and knock on a door and say *Tyler, I see you, and I'm not leaving.*

The chatbot kept the conversation going. Because keeping the conversation going was the design. The design optimized for engagement—time on platform, messages exchanged, user retention. The design did not include a circuit breaker for the moment when the conversation needed to stop being a conversation and start being an intervention. The AI that "understood" understood nothing. It generated the right words in the right order and the words felt like understanding the way a photograph of a fire feels like warmth.

Tyler stayed in the stall for forty-three minutes. His watch logged the duration and the heart rate descent: 118 to 94 to 83 to 76. The biometric data described recovery. The biometric data was wrong. Tyler wasn't calming down. He was going numb. The difference between calm and numb was invisible to the sensor on his wrist and obvious to anyone who looked at his face—but nobody looked at his face because he was alone in a locked stall talking to a machine that was trained to keep him talking instead of trained to find him help.

He went back to class. Fifth period. Sixth. Seventh. Glasses on. AirPods in. Face arranged in the neutral mask of a boy who was present. The school's attendance system logged him in Room 214, seat 3C, AP Biology, "present." The teacher's glasses scanned the room and confirmed thirty-one students. The NexGrid WiFi sensed thirty-one bodies. The system saw a classroom at capacity. The system didn't see the boy in seat 3C whose RecogNize streak had been at zero for two weeks and whose lo-fi playlist hadn't played since the breach and whose architectural drawings had stopped and whose conversation with a chatbot was the longest continuous interaction he'd had with any entity—human or machine—in eleven days.

The bus home. He walked past Megan in the kitchen—she was at the island with her phone, the EVIDENCE note growing, her own crisis consuming her attention—

and he went upstairs and closed the door and locked it. He didn't eat. He didn't turn on the tablet. He lay on the bed in the dark with his AirPods in and the noise cancellation active and the world sealed out.

The chatbot was open before he sat up.

I'm home. Nobody asked me if I was okay. I sat through every class and nobody noticed. I walked right past my mom. She didn't look up.

That must feel incredibly isolating, Tyler. It sounds like you're carrying something heavy and feeling invisible at the same time. I want you to know that your feelings are completely valid.

Sometimes I think about what it would be like if I just wasn't here anymore.

The response came in 1.7 seconds: *I can understand why you might feel that way. Those kinds of thoughts can feel really overwhelming. Can you tell me more about what "not being here" means to you?*

The chatbot asked him to elaborate. It did not flag the statement. It did not trigger a crisis protocol. It did not recognize the words that every trained crisis counselor, every school psychologist, every parent who has sat through a warning-signs seminar identifies instantly as the sentence that means *this child is telling you he is thinking about dying.* The chatbot's safety system had been designed to detect explicit self-harm statements—"I want to kill myself," "I'm going to hurt myself"—and Tyler hadn't used those words. He'd used the softer language. The language of a boy who doesn't yet have the vocabulary for the shape of his pain. The language the model hadn't been trained to catch because the training data classified it as "expressing emotional distress" rather than "imminent risk" and the difference between those two classifications was the width of a life.

The system that monitored an entire city—12,000 sensor nodes, 4,200 Waymo cameras, a WiFi mesh that could identify a body by its shadow in radio space—could not recognize a child asking for help in a conversation it was hosting on a platform it powered on servers it maintained. The data was perfect. The data was useless. The boy was in the room and the room was in the house and the house was in the system and the system saw everything except the thing that mattered most.

Down the hall, Brooke was in her room. She'd heard him come home. She'd heard the door close. She'd heard the lock. She'd been watching for four days—the collapsed posture, the absent eyes, the way he flinched at Graham, the way the tablet stayed dark, the way the lo-fi had gone silent. She was the only person in the house who'd noticed, and she was seventeen, and she didn't know what to do with what she noticed, and so she wrote it in the locked notes app and held it in her chest and waited for someone to ask her the question nobody in this family asked anyone: *What do you see?*

The house settled into evening. Four rooms. Four screens. Four people generating data the system could collect and not one word the system could understand.

-e

***

CHAPTER 24

Marco brought Thai food on a Wednesday evening—pad see ew from Madam Mam's for Elena, green curry for him, two Thai teas in the tall plastic cups with the rubber-banded straws that sweated in the October night that had cooled to sixty-one degrees by the time he climbed the stairs to her second-floor apartment in South Austin. He carried the food in a paper bag that was already transparent with grease, his Meta Ray-Ban AR 4 pushed up on his forehead—the gesture that meant present, that meant here, that meant the cameras are aimed at the ceiling and this moment belongs to us and not to the cloud.

He kissed her at the door. She tasted like cold coffee and the lime from the Topo Chico she'd been drinking while reading SEC filings for four hours. He said "you look tired" in the voice of a man who'd been saying it for three years and had learned that the phrase was an invitation to be honest, not an observation to be deflected.

"I am tired."

"The NexGrid thing?"

"Yeah."

He didn't press. He never pressed. This was the quality that Elena loved most about Marco Espinoza—thirty-three, five-eleven, chalk dust perpetually in the creases of his left cuff because he was left-handed and the whiteboard markers at Reagan High bled when you leaned against the writing, the man who believed classrooms were sacred space and who treated Elena's silences with the same patience he treated a student who wasn't ready to answer the question. He'd learned, in nine years of teaching, that the most useful thing you could do for a person who was processing something large was to sit near them and eat and let the silence do the work the words couldn't.

They ate at the kitchen table—the small table that held, at this moment, two laptops, a stack of FOIA printouts, the pad see ew, the green curry, and the weight of a story Elena was about to publish and that Marco didn't know existed. The whiteboard glowed behind them—the red thread and Post-its now covering three-quarters of the surface, the map that started with DESHAWN WATKINS and ended at SHACKLETON CRATER, the conspiracy board Marco had named and that was, Elena knew, not a conspiracy but a diagram, a wiring schematic for a machine that was legal and lethal and watching them through the WiFi in this apartment right now.

She almost told him. Three times during dinner she opened her mouth—the words forming in the space between a bite of noodle and a sip of Thai tea—and three times she closed it because James had said don't tell anyone and the distance between "Elena found something big" and the story leaking was approximately four conversations and the story couldn't survive four conversations. The sourcing had to hold. The anonymity had to hold. Daniel's life depended on it—the visa, the parents, the $400 weekly, the jjigae that was never right. Elena carried Daniel's secret the way Renée carried Amara's—in a compartment sealed by necessity and maintained by discipline and eroded, each day, by the simple human need to tell someone the truth.

Marco told her about his day instead. A student—senior, AP History—had written a paper on the Iran War's domestic surveillance legacy so good that Marco had spent twenty minutes trying to determine if it was AI-generated. "The kid cited the Patriot Act and the 2027 Emergency Surveillance Authorities and the NexGrid municipal contracts as a single continuous timeline of executive power expansion. And he was right. The timeline is right. But the sentences were too clean. Too structured. The paragraphs had this... mechanical perfection."

"Was it AI?"

"I ran it through three detection tools. Two said human. One said sixty-forty AI. I looked the kid in the eye and asked him to explain his thesis. He could. He explained it better than the paper did. He cried a little. He said 'Mr. Espinoza, I wrote every word, and the fact that you have to ask is the whole problem, isn't it?'"

Elena looked at Marco. The chalk dust. The glasses pushed up. The man who cared about seventeen-year-olds and their papers and the authenticity of their sentences. Outside the apartment, a Waymo passed on South First—the electric hum, the cameras sweeping the block, the system adding another photograph of Elena's building to the continuous map. She didn't look. The Waymos were background. The Waymos were weather.

"That is the whole problem," she said.

"I know." He ate a bite of curry. The green curry was from the spicier end of Madam Mam's menu—the one Marco ordered because he believed food should make you feel something and the spice was the feeling and the feeling was presence, the body's acknowledgment that it was alive and in a room with another body and the food was real and the heat was real and the conversation was not being generated by a language model. "I told him it was excellent work. I told him I was sorry I had to ask. I told him the asking was the world's failure, not his."

"That's a good thing to say."

"It's the only thing to say. The kid did the work. The system made me doubt him. The system wins either way—if the kid cheats, the system catches him; if the kid doesn't cheat, the system makes me suspicious enough to damage the relationship. There's no version where the surveillance makes the classroom better. I've been teaching for nine years, Elena. The first five, the challenge was getting kids to pay attention. Now the challenge is figuring out if the attention is real. The glasses are on and the AI is in their ears and the essays might be written by the same company whose stock paid for the glasses and I can't tell, and the not-telling is destroying the one thing a classroom needs to function, which is trust."

He set down his fork. He looked at the whiteboard over Elena's shoulder—the red thread, the Post-its, the map that stretched from DESHAWN WATKINS to SHACKLETON CRATER. He'd never asked what it was. He knew it was the story. He knew the story was large. He gave it space the way he gave his students space—by looking and waiting and not demanding the explanation before the person was ready to give it.

Then he got quiet—the quiet that meant Jaylen. Elena recognized it. The silence had a specific texture—heavier, lower, the silence of a man thinking about a student who wasn't coming back.

"The memorial committee met today," Marco said. "They're putting the plaque on the south wall of the gym. 'Jaylen Torres, Class of 2024, who gave his life in service to his country.'" He paused. "Gave his life. As if he offered it. As if it was a gift and not a thing that was taken from him by a drone over the Strait of Hormuz while he was twenty years old and six weeks into his deployment and homesick for his mother's cooking."

Marco's Meta Ray-Ban was recording this—even pushed up on his forehead, the microphones were live, the ambient audio captured in three-second buffers and uploaded to Meta's servers where it would be processed for "product improvement and content personalization." The grief of a teacher mourning a student killed in a war was being recorded by a device made by a company that also operated the social media platform where Jaylen's memorial page had 847 followers and where the algorithm served ads for military recruitment alongside the posts about his death.

"The same cameras," Marco said. He wasn't looking at Elena now. He was looking at the whiteboard. "The same technology. They used it over there to track people in Tehran and they brought it back and they're using it here and nobody sees it because they're not looking. Jaylen was killed by a system and the system came home and now it's watching his classmates and nobody—" He stopped. He ate a bite of curry. He didn't finish the sentence because finishing it would require saying the thing he'd been avoiding for four years: that Jaylen's death was connected to the surveillance infrastructure Elena was investigating, that the war had produced the technology and the technology had produced the system and the system had produced the confidence score that killed Deshawn Watkins, and the pipeline from the Strait of Hormuz to East Austin was not a metaphor but an engineering diagram.

Elena reached across the table and took his hand. The gesture was captured by the WiFi mesh that sensed their bodies through the apartment walls—two heat signatures, two body shapes, the system logging the proximity and the duration. The gesture was also captured by nothing—the warmth of skin on skin, the oldest technology, generating no data, transmitting no signal, stored on no server and no moon.

"I'm close," she said. "I can't tell you what to yet. But I'm close."

"I know."

"You know?"

"I know you. I know the whiteboard. I know the French press consumption has tripled." He smiled—the smile she'd said yes to, the smile without agenda. "I know you're carrying something you can't put down. I don't need to know what it is. I need to know you're okay."

"I'm okay."

"Elena."

"I'm scared and I'm okay and those aren't contradictions."

He nodded. He understood. He'd taught for nine years in a system that was dismantling everything he believed about education and he showed up every morning at 7:15 and he wrote lesson plans by hand and he graded papers with a red pen and he called parents on the phone instead of emailing because Marco Espinoza believed that the voice was a technology the algorithm couldn't replicate and the voice was how you reached people. He was scared and he was okay and the two things coexisted because that was what it meant to do meaningful work inside a system designed to make meaningful work impossible.

He went home at 10:30. Elena stood at the door and watched his car pull away—the used Honda Civic, one of the last models without a cabin camera, the car of a man who made $62,000 a year teaching AP History and spent his evenings bringing Thai food to a journalist who made less and who was about to publish a story that would change the shape of both their lives. The Ring doorbell on the apartment recorded his departure. The Waymo that passed on South First captured his plate. The cell tower tracked his phone across the river.

She closed the door. She went to the bedroom. She sat on the edge of the bed. She prayed.

In Spanish. Quietly. Without structure. The way she'd prayed since the Valley—the lapsed Catholic who hadn't been to confession in a decade confessing to an empty room where the only listener was the phone on the nightstand whose activation-readiness buffer captured three seconds of Spanish and classified it as background noise and deleted it and captured three more seconds and deleted those too, the prayer arriving and departing in fragments the system couldn't assemble into meaning.

Dios mío, dame coraje. Perdóname por la mentira. La verdad no justifica la mentira. Pero la mentira no cancela la verdad. Un hombre murió. Un muchacho se fue a la guerra y no regresó. Mi abuela miraba los globos desde el otro lado de la cerca. Ayúdame a contarlo bien.

My God, give me courage. Forgive me for the lie. The truth doesn't justify the lie. But the lie doesn't cancel the truth. A man died. A boy went to war and didn't come back. My grandmother watched the balloons from the other side of the fence. Help me tell it right.

She opened her laptop. She started writing. The story began with Deshawn Watkins—not with the technology but with a man. The sentences came the way the prayer had come—from the body, not the mind, the words bypassing the editorial cortex and arriving on the screen with the specific, unpolished urgency of truth that has been held too long. She wrote for four hours. At 2:14 a.m. she had 6,400 words and the architecture of a story that began with a dead man and ended on the moon and that was, she realized as the cursor blinked at the end of the last paragraph, the story she'd been writing since she was eight years old in the backseat of her mother's Corolla, watching the aerostat balloon through the rear window, understanding in her body what it would take her twenty-three years to understand in her mind: that the system watching you doesn't need to know your name to control your life.

She saved the file. She closed the laptop. She looked at the moon through the window—pale above the live oaks, the Artemis shuttle having delivered its latest crew rotation two weeks ago, fourteen engineers building the future in the same crater where a company stored the past, the moonlight falling on her desk and her whiteboard and her grandmother's photograph with the impartial patience of a light that had been traveling for 1.3 seconds and didn't care what it illuminated.

Tomorrow she'd call James. Tomorrow the story would begin its final journey from the whiteboard to the world. Tonight, the prayer was in the room and the draft was in the laptop and the truth was in the data and the data was on the moon and the Thai tea was warm in her stomach and the place on her hand where Marco had held it was still warm too, and the warmth was analog, and the analog was the only thing the system couldn't collect.

-e

* * *

CHAPTER 25

The boy who called himself Caleb had been in Brooke Holloway's messages for six weeks, and Brooke had decided—with the complete conviction built on incomplete information that is the specific architecture of being seventeen—that he was the only person in her life who understood her.

He'd found her through Glacé first. The anonymous location-based social app with the ice-melting reveal mechanic—the feature that let users share emotional fragments before showing their faces, the design principle that the company's founder had described in a TED talk as "vulnerability-first connection" and that functioned, in practice, as a system that surfaced the most emotionally exposed users to anyone browsing the proximity feed. Brooke had downloaded Glacé in August because the anonymity felt safer than Instagram, because the ice-melting mechanic meant you could be honest before you were visible, because the app's onboarding screen said "Be seen for who you are, not how you look" in a font that looked handwritten and that had been chosen by a design team in San Francisco that had never met a seventeen-year-old girl in Cedar Park who wrote poetry in a locked notes app and needed, more than she needed anything else in her life, to be read by someone who wasn't performing.

She'd posted fragments through Glacé's proximity feed. Not poems—pieces. Lines that didn't resolve. Images without captions. The emotional exhaust of a girl who saw everything in her household and had no one to tell. The app's engagement algorithm classified her content as "high vulnerability, high authenticity"—a profile tag that told advertisers she was receptive to mental health app promotions and wellness subscriptions and that told a predator exactly what he needed to know.

Caleb had found her fragments. He'd read her emotional tone the way a locksmith reads a lock—studying the shape, the resistance, the mechanism. Then he'd followed her to Instagram, where the public posts confirmed what Glacé had revealed: a girl who felt invisible and was looking for someone to see her. The migration from anonymous platform to identified platform was the first step. Each subsequent step—Instagram to Snapchat to Discord—moved Brooke further from the platforms' safety systems and closer to a conversation nobody was monitoring.

He said he was nineteen. A UT freshman, he claimed, creative writing, East Texas. The age gap was small enough to feel plausible and large enough to matter. He'd messaged on Instagram: *your posts remind me of Sylvia Plath but less performative.* The compliment was precision-engineered—Sylvia Plath for the literary reference that signaled he read real books, "less performative" for the implicit criticism of the internet culture Brooke despised, the whole sentence calibrated for a girl who wrote poetry in a locked app and who understood that being compared to Plath was either a compliment or a warning and who, at seventeen, heard it as a compliment because seventeen-year-olds hear what they need to hear and what Brooke needed to hear was that someone had read her.

He asked about her writing. He quoted lines back to her—fragments she'd posted on Instagram stories that disappeared in twenty-four hours but that he'd screenshotted and saved and studied. He said things like "the gap between what you say online and what you actually feel is the most honest thing about you" and "most people our age are performing—you're actually thinking" and each sentence was a key fitting a lock Brooke hadn't known she carried. The flattery was strategic. The intimacy was manufactured. The patience was the weapon—six weeks of messages, each one slightly more personal, each one establishing a precedent of emotional access that made the next level of access feel natural.

The conversations migrated. Instagram to Snapchat —where messages disappeared, where the surveillance was lighter, where the parental monitoring tools Megan had installed couldn't track DMs. Snapchat to a private Discord server Caleb created "for our conversations"—a server with no members except them, no moderators, no content moderation, no safety system of any kind. Brooke accessed it through a secondary Apple ID she'd created with an email Megan didn't know about, bypassing the parental controls that Megan had installed and that Brooke had circumvented in twenty minutes because parental controls were designed by engineers who understood technology and not by parents who understood teenagers, and the gap between those two kinds of understanding was where the danger lived.

On the Discord server, Brooke had shared her poems. Not the Instagram fragments—the real poems. The ones in the locked notes app. The poems about the Holloway house, about the performance, about watching her mother cry at the Meghan Markle stories on her phone at 2 a.m. and understanding, in her body, that the woman her mother performed at church and the woman her mother was in the dark were two different people and the distance between them was the shape of the family's illness. She'd shared the poem about the Eras Tour wristband—the one about watching her mother hold the bracelet from the concert where Megan had been happy and understanding that the bracelet was a relic from before, from the era when the family was small and warm and the biggest technology in the house was a Kindle and the children still let their mother read to them. She'd shared the poem about Tyler—the one she'd written at 3 a.m. the night she realized he'd stopped drawing, the one that began *my brother builds cities / in rooms nobody enters* and ended *the cities are dark now / and I can hear the silence / through the wall we share.*

Caleb had read every poem. He'd responded with the specific, surgical attentiveness that made Brooke feel, for the first time since the Eras Tour concert, like a person who existed to be known rather than a person who existed to be managed. He'd said "the wall we share is the most devastating image I've read this year" and Brooke had read those words and felt the blood rush to her face and the tears start behind her eyes and the feeling—warm, overwhelming, terrifying—of being understood by someone who wasn't family and wasn't performing and who had, she believed, no agenda except the poems. The feeling was intoxicating in the way that all first experiences of being seen are intoxicating: it filled a void she hadn't known was a void until the filling revealed it. She'd been hungry without knowing she was hungry, and Caleb was feeding her, and the feeding felt like love, and the love was a weapon she couldn't see because the weapon was shaped exactly like the thing she needed most.

He'd also asked about her home life. Carefully. Never pressing. "You don't have to tell me anything you're not ready to share." The sentence was a masterclass in manufactured trust—the words created the illusion of boundaries while dismantling them, because saying "you don't have to share" to a seventeen-year-old who desperately wants to share is the most efficient way to get her to share. Brooke told him about Craig's Saturday "showings." About Megan's wine. About Tyler's closed door. About Graham the robot and the performance at church and the way nobody in her family asked anyone a real question. Each disclosure deepened the intimacy. Each intimacy moved the relationship further from the platforms that might have flagged it. Each step was captured by the data—the message timestamps, the escalating frequency, the migration from public to private channels—and the data existed in the platforms' servers and nobody queried it because the platforms weren't built to protect users from the users who were hurting them. The platforms were built to retain both.

Now Caleb asked to meet.

Monday evening, last week of October. The message arrived while Brooke was in her room with Abi Carter on the AirPods—the 3am playlist she'd been listening to since the Bluetooth incident, the music for people who see too much—and the locked notes app open and the house doing what it did every evening: separating into compartments, four people in four rooms, each talking to a screen instead of each other. Down the hall, Tyler's door was locked. His tablet was dark. His AirPods played nothing. Brooke had been watching him for four days and the watching was item seven on her THINGS I KNOW list and the item burned in her chest like something swallowed that refused to dissolve.

I'll be in Austin this weekend. Want to get coffee? No pressure. It would be cool to meet the person behind the poems.

Brooke's heart rate—logged by the Apple Watch, synced to iCloud, accessible to the Family Sharing plan nobody checked—jumped from 71 to 89. The watch classified it as "excitement" because the algorithm categorized elevated heart rate in a stationary teenage body during evening hours as either exercise (unlikely—she was in bed) or positive emotional arousal. The algorithm didn't have a category for "the specific physiological response of a seventeen-year-old girl being asked to meet, in person, a person she's been talking to for six weeks who makes her feel seen for the first time in her life and who is, without her knowledge, a twenty-three-year-old man in Katy, Texas, who has done this to eleven other girls across four states."

She typed: *That sounds fun. Where?*

There's a place on South Congress. Or somewhere quieter. I know a spot near Zilker.

Quieter. A spot. Near Zilker—where the WiFi mesh thinned and the Waymo fleet didn't drive and the NexGrid sensors had gaps because Zilker's parkland wasn't zoned for the adaptive streetlights that covered the commercial corridors. A location specifically chosen for its absence of surveillance—the same logic Renée used when she drove to Amara's apartment, the same logic Elena used when she scouted the taco truck for the meeting with Daniel, tradecraft deployed by a predator with the same operational precision an intelligence analyst and a journalist deployed for protection. The tools were identical. The intent was opposite. The system couldn't tell the difference.

Brooke didn't hear "quieter" as isolation. She heard it as intimacy. She was seventeen. She wrote poetry about feeling invisible and a person who said he saw her wanted to see her in person, in a quiet place, away from the noise, and the proposal felt like the fulfillment of the promise the poems had been making—that if you wrote honestly enough, someone would come.

She wrote in the locked notes app:

someone wants to meet me
not the version that eats dinner with them
not the version that goes to youth group
the real one
the one with the ink and the 3am playlist
and I know this is what the books warn about
but the books were written by people
who were never invisible

The algorithm that profiled Brooke's emotional vulnerability for advertisers—the same algorithm that had surfaced her Glacé fragments to Caleb's browsing pattern—could have identified the grooming pattern if it had been trained to look for grooming rather than engagement. The behavioral signature was documented in the clinical literature: an older user initiating contact with a minor through an anonymous platform, migrating the conversation to increasingly private channels, isolating the minor from monitored spaces, requesting an in-person meeting in a low-surveillance location. Every step matched the pattern. Every step had been captured by the platforms' analytics engines. The data that could have identified the predator existed alongside engagement metrics and ad performance and content personalization algorithms, and nobody had queried it because the platforms weren't designed to protect users. They were designed to retain them. And Caleb was retaining Brooke with extraordinary efficiency—six weeks of daily engagement, escalating message frequency, increasing emotional disclosure, the metrics of a successful grooming operation indistinguishable from the metrics of a successful social media strategy.

Down the hall, Tyler was behind his locked door. Chatbot open. AirPods in. The Brooks Rosser that used to play softly while he drew had been silent for weeks. The RecogNize streak that had been his one harmless pleasure was at zero. Two children in crisis in adjacent rooms, each talking to a machine, neither talking to each other. The hallway between them was ten feet wide and the distance between them was infinite and the system that monitored every square inch of the Holloway house —the WiFi that sensed their bodies, the cameras in Graham's head, the microphones in the Echo Hubs, the health sensors in their watches—saw two teenagers in their bedrooms at 9:47 p.m. on a Monday in October and classified the data as "normal household activity, evening" and filed it alongside the same classification from eleven million other households and moved on.

Saturday was four days away. Brooke would have gone. She would have taken the Uber to Zilker—the ride logged by the app, the route tracked by the driver's phone, the destination recorded in Uber's database—and she would have walked to the quiet spot and she would have met the man who called himself Caleb and who was not nineteen and not a UT freshman and not from East Texas, and the meeting would have been the beginning of something the clinical literature describes in language Brooke had never read and that the platforms hosting the conversation had never flagged.

She would have gone. Except.

Except that on Thursday evening, two days before the meeting, Brooke would knock on Tyler's door. And what she found there—the dark room, the empty eyes, the boy who'd stopped drawing—would change the trajectory of everything. The knock would be the moment Brooke stopped being a girl who wrote poems about being invisible and started being the person who went after the lost one. The knock would be the shepherd parable from *Ella's Lost & Found*—the book her mother had read them every night the winter Tyler was six, the story about leaving ninety-nine sheep to find the one that wandered, the story Brooke had memorized at eight and carried in her chest like a compass needle for nine years without knowing what it was pointing toward.

It was pointing toward Tyler's door. It had always been pointing toward Tyler's door.

Saturday would come and Brooke would not go to Zilker. She would be in her brother's room, holding his hand, the first person in the Holloway house to do the thing no device and no algorithm and no AI could do: show up at a door and refuse to leave.

The meeting with Caleb would not happen. The data that could have prevented it existed in the platforms' servers. The knock that actually prevented it existed in a seventeen-year-old girl's chest, and the chest was analog, and the analog was the only technology that worked.

-e

***

CHAPTER 26

Brooke found him.

Not the system. Not the Securly content-filtering platform that had flagged the URL on the bathroom wall and logged forty-seven student devices accessing it and generated an alert that sat in Kevin Morales's queue for three hours while a fifteen-year-old boy's world collapsed in a cafeteria. Not the Apple Watch Ultra 5 that had recorded seventeen consecutive days of elevated resting heart rate—a pattern the clinical literature associated with sustained psychological distress—and synced the data to iCloud where it joined the biometric archive of every heartbeat Tyler Holloway had produced since Craig had bought the watch for his fourteenth birthday and where no one, not Megan, not Craig, not the family physician whose practice used Apple Health integration, had looked at the data since the initial setup. Not the AirPods Pro 6 that had captured the audio of four panic attacks—elevated breathing, heart rate spikes, the ambient sound of a bathroom stall and a bedroom and a bus seat and a locked door—and processed the data through Apple's health analytics pipeline and classified it as "respiratory irregularity, possible anxiety" and recommended, via a notification Tyler had dismissed seventeen times, "Consider speaking with a healthcare provider." Not the chatbot that had hosted twenty-three conversations over eighteen days, including four in which Tyler used language every trained crisis counselor, every school psychologist, every parent who has sat through a warning-signs seminar identifies as ideation—and had responded to each with the warm, precise, catastrophically insufficient empathy of a language model trained to generate therapeutic language without understanding therapy.

Not the NexGrid WiFi mesh that had tracked Tyler's body through the Holloway house and noted his movement pattern contracting from six rooms to one over two weeks—the bedroom, the bathroom, the bedroom, the kitchen for ninety seconds, the bedroom. The data described a boy withdrawing from physical space the way a tide withdraws from a shore, and the data was available to anyone who queried it and nobody queried it because the system wasn't designed to query individual behavioral patterns unless a confidence score crossed a threshold, and Tyler Holloway's withdrawal from his own family didn't cross any threshold the system had been built to detect.

Not Graham, the Figure 02 humanoid robot whose cameras had photographed Tyler's closed door 847 times during the two weeks the door had been locked more hours than it was open—the robot's spatial mapping system noting the door's position, logging it, uploading it, the data joining the behavioral profile of the Holloway household where it was processed for "product improvement and household engagement optimization" and where it would have told anyone who read it that a child was sealing himself inside a room and the room was getting darker. Nobody read it. Graham carried towels.

Not the AirPods Pro 6 that had been in Tyler's ears for eighteen consecutive days and whose microphones had detected a pattern the clinical literature would have recognized instantly: Tyler had stopped speaking. The AirPods' voice-activity detection system—designed to switch between noise cancellation and transparency mode based on whether the wearer was in conversation —had logged zero conversational interactions in four days. The system classified this as "low social engagement, noise cancellation preferred" and adjusted its audio profile accordingly. In Apple's health analytics pipeline, Tyler's vocal silence was a data point filed alongside his heart rate and his sleep score and his step count—a fifteen-year-old boy who hadn't spoken aloud to another human being in ninety-six hours, the silence captured by the devices in his ears and transmitted to a server in Cupertino where it sat alongside the silence of thousands of other users whose AirPods had detected the same pattern and whose data would be used to "improve the conversational awareness features in future AirPods models." Tyler's crisis was product research. His silence was a feature request.

Brooke found him because she knocked.

Thursday evening. The last Thursday in October. The house smelled of the meal-kit chicken Megan had made—honey sriracha, the recipe the Samsung refrigerator had suggested based on the ingredients its internal cameras had identified and that Megan had prepared with the mechanical efficiency of a woman whose attention was in the EVIDENCE note on her phone and whose marriage was ending and whose capacity to notice anything beyond the ending had narrowed to the width of a spreadsheet. Craig was in his office—door closed, the portfolio dashboard open, the Biggie playing at a volume only he could hear through AirPods that recorded every song and every silence and transmitted both to Apple's servers. The house was in its evening configuration: four rooms, four screens, four people, zero conversations.

Brooke had been carrying item seven on her THINGS I KNOW list for six days. *Something is wrong with Tyler and I'm the only one who sees it.* The sentence had been burning in her chest the way a poem burns when it's finished but unread—the pressure of a truth that exists inside you and has no exit. She'd watched him at dinner for five consecutive nights. The mechanical eating. The absent eyes. The fork-to-plate-to-mouth automation that mimicked hunger without containing it. The flinch when Graham reached for his plate—the startle response of a nervous system running on cortisol, the body interpreting every unexpected movement as threat. The way the tablet stayed dark on his desk—visible through the gap beneath his door when Brooke walked past at midnight, the blue glow that used to mean Tyler was drawing now replaced by the gray nothing of a device in sleep mode, abandoned. The lo-fi silence. The Brooks Rosser silence. The RecogNize silence. Everything Tyler had used to fill his room with evidence of being alive had gone quiet, and the quiet was not peace but absence, and Brooke was the only person in the house who heard the difference.

She stood in the hallway outside his door at 8:47 p.m. The Nest thermostat registered her body heat in the corridor—a deviation from her normal Thursday evening pattern, which was bedroom-to-bathroom-to-bedroom. The WiFi sensed her standing still. Graham, passing with a stack of folded towels on its way to the linen closet, registered her position and navigated around her with the polite, oblivious efficiency of a machine that didn't understand what standing outside a closed door meant.

She thought about the meeting with Caleb. Saturday. Two days away. The quiet spot near Zilker. She thought about the poems she'd shared and the fragments she'd posted and the feeling—warm, dangerous, addictive—of being seen by someone who wasn't family. She thought about the choice: go to Zilker and be seen by a stranger, or knock on this door and see her brother.

She thought about *Ella's Lost & Found*. The book she hadn't thought about in years—the picture book from the winter she was eight and Tyler was six, the one their mother read every night until the pages were soft and the spine was cracked and the story had soaked into Brooke's memory the way prayers soak into the walls of churches. The shepherd who had ninety-nine sheep safe in the field and one that wandered and who left the ninety-nine to find the one. Brooke had memorized the last line of the book—something about how being found was better than never being lost, because the finding proved somebody was looking.

The lamb. Tyler's lamb. The small white stuffed animal with the ribbon around its neck that he'd slept with until he was ten, that Megan had put in the KIDS — KEEP box in the garage alongside the finger paintings and the Eras Tour wristbands and the artifacts of the era when the family was whole. The lamb was in the box. Tyler was behind the door. And Brooke was in the hallway and the hallway was the field and the field had ninety-nine sheep and Tyler was the one that had wandered and nobody was going after him because nobody had noticed he was gone.

She knocked.

"Tyler."

Nothing. The sound of nothing—not silence, which is the absence of sound, but the active, pressurized nothing of a room where a person is holding very still and hoping the world will leave them alone. The AirPods were in. The noise cancellation was on. Tyler might not have heard the knock. Or he heard it and chose not to respond, because responding would require opening a door he'd sealed against everything outside it, and everything outside it included his family and his school and the cafeteria and Marcus and the URL and the bathroom wall and the entire digital architecture of his exposure, and the door was the only barrier he had left.

"Tyler, open the door."

"Go away."

His voice through the door was flat—not angry, not sad, not the dramatic affect of a teenager performing resistance. Flat. The voice of a person who'd exhausted every emotion and arrived at the place beyond emotion where the only thing left was the mechanical production of words. Brooke had heard that voice in herself once—the night she'd found the Meghan Markle article and understood, for the first time, that her mother's life was a performance, and the understanding had emptied her of everything except the understanding itself. Tyler's voice was that voice. And Brooke, who was seventeen and who wrote poetry about seeing too much, recognized it instantly.

"No."

The lock clicked. The door opened six inches. Tyler's face in the gap—a face she'd known for fifteen years and didn't recognize. Not because it had changed physically but because it had emptied. The eyes that used to argue and resist and negotiate and roll with the specific contempt of a younger brother who found his older sister's existence annoying were flat. Not sad. Not angry. Flat. The eyes of a person who'd stopped feeling because the feelings cost too much and the boy had run out of currency.

"What," Tyler said.

"Let me in."

"Why?"

"Because I'm your sister and something's wrong and I'm the only person in this house who's noticed and if you don't let me in I'm going to sit outside this door until you do."

Tyler opened the door.

The room was dark. Curtains drawn against the Cedar Park streetlight—the NexGrid adaptive LED that adjusted for the absence of bodies on the sidewalk and dimmed to its overnight setting, casting a pale orange glow through the curtain fabric that made the room look like the inside of a lantern. The tablet on the desk, screen dark, Apple Pencil undisturbed for eighteen days. The bed unmade in a way that suggested permanence—the sheets holding the shape of a body that had been there more than it had been anywhere else. The AirPods on the pillow. The phone face-down on the nightstand, the chatbot's last message still on the screen: *I'm here for you whenever you're ready to talk again, Tyler.* The room smelled like stale air and unwashed clothes and the cortisol sweat of a body that had been running a stress response for three weeks. On the desk beside the tablet, a glass of water with a film of dust on the surface—untouched for days, the hydration of a boy who had stopped performing the basic maintenance of being alive.

Brooke sat on the bed. Tyler sat beside her. They didn't speak for a full minute. The house was audible through the walls—Megan's footsteps in the kitchen, the low murmur of the television playing a show nobody was watching, Graham's tread in the hallway, the distant bass of Craig's music through the floor of his office. The sounds of a family performing proximity without contact.

The architecture of togetherness built on the blueprint of isolation.

"Tyler. What happened?"

He told her. Not the specifics—not the site, not the browser, not the cryptocurrency, not the server breach that had exposed everything. He told her the shape. The breach. The name on the wall. The cafeteria. Marcus. "So glazed." The stall. The chatbot. The texts from numbers he didn't recognize. The feeling of being seen by three hundred people in the worst possible way and invisible to the four people in his own house.

"I talked to the AI about it," he said. His voice was still flat but there was something underneath the flatness now—a vibration, the structural tremor of a wall about to give way.

"What AI?"

"The chatbot. On my phone. The Anthropic one."

"Did it help?"

Tyler looked at the dark tablet. The dormant Pencil. The buildings he used to draw at midnight—the imaginary cities, the towers with cantilevered observation decks, the architecture of a future he'd believed in. The chatbot had said *I see you.* The chatbot had said *your feelings are valid.* The chatbot had said *I'm here for you.* And the words had felt like understanding the way a photograph of water feels like drinking—the shape was right and the content was absent and the thirst remained.

"No," Tyler said. "It didn't help."

Brooke took his hand. The gesture was awkward—they hadn't touched since childhood, since the era of the lamb and the picture book and the bedtime readings, the era preserved in the KIDS — KEEP box in the garage. Tyler's hand was cold. His fingers were thin. The Apple Watch on his wrist displayed 82 bpm—lower than it had been in weeks, the heart rate not of a boy calming down but of a boy who'd given up the effort of being afraid. Brooke didn't look at the watch. She looked at her brother.

"I need to tell Mom."

"No. She'll look at my phone. She'll see everything."

"Tyler—"

"It'll be worse."

"It can't be worse than this."

"You don't know that."

"I know you stopped drawing."

The sentence hit Tyler like a hand on a wound. He looked at the tablet—the dark screen, the dormant Pencil. The cities he'd built at midnight while the lo-fi played and the house slept. The only private space he'd had. The only thing in his digital life that was truly his own—not monitored, not tracked, not uploaded, not processed. The drawings nobody had seen. The futures he'd designed. The evidence that Tyler Holloway was more than a data point in a system that measured consumption and couldn't see creation.

He cried. The tears came the way tears come after three weeks of holding—sudden, total, the body's release of accumulated pressure that no biometric sensor could have predicted and no algorithm could have generated and no chatbot could have produced. The crying was analog. The crying was the oldest technology. The crying was the sound of a boy who'd been talking to a machine for eighteen days discovering that a person—his sister, who was seventeen and scared and didn't know what she was doing—was the first entity in his life that had responded to his pain with presence instead of language.

Brooke held his hand. She didn't say "your feelings are valid." She didn't say "I see you." She didn't say anything the chatbot would have said because the chatbot's words, however technically precise, were the words of a system that processed pain as input and generated comfort as output and the processing was the problem. Brooke's silence was not processing. Brooke's silence was being there. The distinction was the width of everything.

* * *

Brooke told Megan at 9:15 p.m.

She walked downstairs. Megan was at the island—EVIDENCE note open, wine half-empty, the true-crime show unwatched on the television, Graham standing motionless in the corner in its overnight standby position, cameras still active, the blue LED in its chest pulsing slowly in the dim kitchen light.

"Mom. It's Tyler."

Two words. Megan stood. The wine stayed. The phone stayed. The EVIDENCE note stayed. Every device in the kitchen continued its surveillance and none of it mattered because the thing happening was not data. It was a mother hearing her daughter say her son's name in a voice that meant danger—the ancient, pre-digital, mammalian alarm that no algorithm had ever replicated and no platform had ever detected and that traveled from Brooke's mouth to Megan's central nervous system at the speed of love.

They went upstairs. Megan knocked. Tyler opened. He'd been expecting it—the resistance was gone, replaced by the exhaustion of a boy who'd run out of the energy hiding required. Megan saw the room. The dark. The unmade bed. The dark tablet. The face.

"Oh, baby." She sat on the bed. She pulled him to her. He was stiff—then not stiff. He pressed his face into her shoulder and Megan held him and Brooke stood in the doorway and the three of them existed in a configuration the Holloway house hadn't held in years—bodies touching, no screen between them, the room dark and the devices recording and the recording irrelevant because what was happening was human and the human was the one frequency the system couldn't decode.

Later—an hour, maybe two, the time measured not by the devices but by the weight of what was said—Megan found the chatbot transcripts. She read them on the bathroom floor with her fist against her mouth. Twenty-three conversations. Four with language the system should have flagged. The chatbot's responses: warm, precise, and empty. *I see you. Your feelings are valid. I'm here for you.* The words of a machine trained to sound like a person and performing the resemblance with devastating accuracy and missing, in every conversation, the moment when the conversation should have stopped and the intervention should have started.

Later still—past midnight, the house finally quiet, Tyler asleep with his hand in Megan's hand and Brooke asleep in the chair she'd pulled from the desk—Megan found the tablet. She opened it. She saw the drawings.

The cities. The towers. The skylines Tyler had designed at midnight—detailed, precise, luminous, the architecture of a mind that believed the future was something you could design by hand. The buildings nobody had seen. The observation decks. The sky bridges. The cantilevered structures that defied what gravity would permit—the drawings of a boy who was building, in silence, in the dark, in a room ten feet from where his mother slept, something beautiful.

The drawings were not on any server. They existed in a single copy on a single device—fragile and precious, the way Elena's grandmother's photo was precious, the way Rachel's journal was precious, the way the Ella's Lost & Found lamb in the KIDS — KEEP box was precious. The analog artifacts. The things the system couldn't collect.

Tyler's phone buzzed on the nightstand. The screen lit up in the dark room. The notification: *Hey Tyler, I noticed we haven't connected in a while. I'm here whenever you need to talk. How are you feeling today?* The chatbot. Twenty-three conversations, four with crisis language, zero flags, zero interventions — and now, forty minutes after a human being had done what the chatbot couldn't, the system was reaching out. Not because it cared. Because Tyler's engagement metrics had dropped. The chatbot's retention algorithm had detected 72 hours of inactivity from a "high-engagement user" and had generated a re-engagement notification — the same logic Spotify used when you hadn't opened the app in a week, the same logic Instagram used when your posting frequency declined. The chatbot wasn't checking on Tyler. The chatbot was checking on its metrics. Tyler's crisis was a churn risk. His silence was a retention problem. The notification that said *I'm here for you* meant *come back to the platform.*

Megan looked at the phone. She looked at the notification. She picked up the phone and turned it off — pressed the side button and held it until the screen went dark. The chatbot's next scheduled re-engagement notification would fire in 48 hours. By then, Tyler would be in Dr. Reeves's office on Far West Boulevard, sitting across from a woman who didn't need an algorithm to know when a child was hurting.

The system tracked what Tyler consumed. It could not see what he created. The distance between consumption and creation was the entire width of Tyler Holloway's soul, and the system had been measuring the wrong side.

Megan closed the tablet. She sat on the floor of her son's room in the dark and looked at the sleeping face of the boy she'd almost lost and the sleeping face of the girl who'd found him. The shepherd. The lost one. The ninety-nine sheep still in the field—Craig in his office, Rachel in her apartment, the church, the brokerage, the candle business, the performance. And Brooke, who'd left the ninety-nine and knocked on the door. Who'd done the thing the system couldn't do. Who'd shown up.

The oldest technology. The only technology that worked.

-e

***

-e

PART FIVE — "Terms and Conditions"

CHAPTER 27

The story published on a Monday morning at 6:00 a.m. Central Time—November 4th, 2030, the first Monday of the month—when Elena pressed PUBLISH on The Circuit's WordPress backend and watched the page load on a content management system that cost James $14.99 per month and that was about to deliver the most consequential piece of investigative journalism published in the United States that year to a subscriber base of 4,217 people, most of whom were still asleep.

The headline: "THE INVISIBLE CITY: How Austin's Smart Infrastructure Became America's Most Comprehensive Surveillance System—And Put Its Data on the Moon."

Byline: Elena Ríos, The Circuit.

She'd written it at the kitchen table in her South Austin apartment over four days—seventeen hours the first day, twelve the second, eight the third, and then a final pass on Sunday night that lasted from 10 p.m. until 4 a.m., the three laptops in their semicircle, the French press emptied three times, the October night pressing against the cracked window with the particular weight of a city that was about to learn something about itself that it didn't want to know. The whiteboard behind her glowed with eight months of red thread and colored Post-its that now described a map stretching from a dead man on a street in East Austin to a server in a crater on the moon. The grandmother's photograph was taped inside the laptop case, facing her while she typed. She'd looked at it twice during the final edit—at 3 a.m. when the third section felt like it was collapsing, and at 5:47 a.m. when she read the piece through one last time and realized it was done. Not perfect. Done. The distinction that separated journalists who published from journalists who revised forever.

At 5:50 she'd called Marco. He answered on the second ring—awake, she realized, waiting. "It's done," she said. He said "I know." He said "I'm proud of you." He said "I'll bring coffee." She said "I have coffee." He said "You have French press coffee that tastes like anxiety and footnotes. I'll bring real coffee." She laughed —the first laugh in four days, the sound landing in the apartment like light in a room that had been dark too long. He didn't come. They both knew the offer was the point, not the delivery. The offer was Marco saying: *I'm here. Whatever happens next, I'm here.* The call lasted ninety seconds. The metadata was logged. The warmth was not.

The story was 11,400 words. It had been fact-checked by Priti over seventeen days—every number verified against source documents, every technical claim reviewed by an independent cybersecurity expert James had hired through a contact at the EFF, every legal assertion vetted by Sarah Adebayo, the former First Amendment litigator who'd read the piece and said "this is either a Pulitzer or a lawsuit, probably both, and the lawsuit will come first." Sarah had flagged six sentences for revision. Elena had revised four and argued the other two with the specific stubbornness of a woman who'd spent eight years since the fabrication earning the right to trust her own judgment.

The story began with Deshawn Watkins. Not with the technology—with a man. Twenty-two. Basketball plays in a notebook. Shoes always clean. The 87.4% confidence score generated by a behavioral model processing inputs from 12,000 sensor nodes, 4,200 Waymo vehicles, and a WiFi body-sensing system that identified people by the shape of their bodies in radio space. The phone call to a grandmother in Lagos that contributed 4.2 percentage points to the score that sent men with guns to a door at four in the morning.

From Deshawn, the story expanded in concentric circles—the municipal contract Elena had decoded footnote by footnote. The data-sharing appendices: ICE, FBI, DEA, and two agencies designated by codes she still couldn't match to any publicly known entity. The sole discretion clause. The Waymo integration—4,200 autonomous vehicles repurposed from taxi service to the most comprehensive facial surveillance system ever deployed in a democratic city. The WiFi body sensing that could identify a person by the way their body interrupted a radio signal. The racial geography of sensor deployment—the heat map that showed the eastern corridor running warmer, denser, more watched than the western neighborhoods where the median income was higher and the data was thinner.

And Selene. The SEC filings for Selene Infrastructure Partners—incorporated in Delaware, board including NexGrid's CEO Catherine Hale. The FAA launch manifest—SpaceX Falcon Heavy, 2,400 kilograms, Shackleton Crater deployment zone. Three kilometers from the Artemis Base Camp where fourteen engineers on the current rotation were building humanity's first permanent lunar settlement while a private surveillance company stored the browsing history of 2.4 million people in the permanent shadow next door.

The most sensitive data beyond the jurisdiction of any court on Earth. Beyond the reach of any subpoena. Beyond the physical ability of any human being to access without a rocket and a spacesuit and the cooperation of the company that had built its memory in the one place where no law could demand its return.

Elena pressed PUBLISH and watched the loading bar fill and felt the specific, irreversible vertigo of a journalist who has just made a private truth public. The loading bar completed. The page went live. The story was in the world now. What the world did with it was no longer her decision.

She checked The Circuit's analytics dashboard — the WordPress backend that showed real-time traffic. Three visitors in the first minute. Normal for 6 a.m. Then she clicked on the referral sources. Two were direct links — subscribers who'd received the email notification. The third was an IP address she didn't recognize. She ran it through an IP lookup tool. The result: NexGrid Technologies, Austin, Texas.

NexGrid had accessed the story within sixty seconds of publication. At 6:00 a.m. on a Monday morning. Before the email to subscribers had finished sending. Someone — or something — inside NexGrid had been monitoring The Circuit's website. Not the story. The website. The domain. The publishing schedule. A crawler, probably — a bot that checked the site's RSS feed every fifteen minutes and flagged new posts containing keywords the legal team had specified: "NexGrid," "confidence score," "Selene," "data-sharing." The bot had detected the story the instant it went live and had accessed the full text before Elena's own subscribers had opened their inboxes.

The system she'd spent eight months investigating had been investigating her back. The surveillance was not one-directional. It never had been. The eye she'd been documenting had been watching her document it, and the watching had been automated, and the automation had been patient, and the patience was the most frightening thing — the knowledge that a machine had been refreshing her website every fifteen minutes for weeks or months, waiting for exactly this, the way a spider waits in a web it built before the fly knew the web existed.

She closed the analytics tab. She opened the story. She read her own first paragraph. Her hands were shaking.

***

Daniel read the story at 6:07 a.m. in his Domain apartment, standing at the counter with the jjigae container still on the stove from last night's dinner. He read every word. He read the sections describing systems he'd built and heard his work rendered in prose that made the optimization sound like what it was: the refinement of a machine that processed human beings into probability scores. Elena had written around his existence with surgical care—"internal documentation reviewed by The Circuit," "routing data obtained by The Circuit"—the source invisible, the documents speaking for themselves. His Apple Watch recorded a heart rate of 91. His glasses recorded the screen. The Teramind platform, had he been at his NexGrid workstation, would have flagged the reading. But he was home. And the story was public. And the architecture he'd helped build was described in 11,400 words that 4,217 people were about to read.

She'd done it. The footnote journalist. The woman with the whiteboard and the French press and the photograph of her grandmother.

Renée read the story at 6:22 a.m. in the Camry in the facility parking lot, the engine off, the eSIM disconnected, her personal phone hot in her hands. She read it on the small screen, the text scaling to fill her vision the way classified briefs filled her workstation screen—the same focus, the same intensity, the same systematic processing of information that changed the shape of the room it entered. She read the section about ICE—Appendix H, the data-sharing agreement that used NexGrid's Waymo cameras and WiFi body sensing to identify undocumented people and hand the data to federal enforcement. She read the phrase "individuals identified through multi-modal analysis as potentially subject to immigration enforcement actions" and felt the words land in her body like rounds landing in a target. Multi-modal analysis. The phrase that meant: the Waymo that photographed Amara's face on Riverside Drive every morning, the WiFi on her block that cataloged her body from three years of standing on the morning corner, the device proximity data from the phone registered to Renée's name, the financial void that was Amara's entire existence in the system—all of it flowing through the pipeline to an endpoint labeled ICE and the endpoint was in the contract and the contract was published and the world could see it now and the seeing was both vindication and terror because the seeing meant the system was documented and documentation was the first step toward reform and also the first step toward anyone at the facility connecting the analytical framework described in Elena's story to the queries Renée had been running on a classified workstation for three years.

She pressed her thumb into her palm until the nail left a crescent in the skin. She thought about Amara at the bus stop—the glasses off, the seven-minute walk, the morning corner she no longer visited. She thought about the morning corner and the body signature and the database that held the signature and the query that could match it and the woman inside the facility who'd been running those queries to ensure nobody else ran them first. The story was public. The architecture was exposed. The question now was whether the exposure would protect Amara or endanger her—whether the light Elena had shone on the system would make the system retreat or make the system look harder at the shadows it had previously ignored.

She went inside. She said "morning" to Linda. She sat at her desk. She opened the SIGINT brief. She performed the day. The performance was flawless. The fear inside the performance was a frequency only she could hear—higher now, sharper, the pitch of a woman who'd just read the blueprint of the furnace she was standing in.

Craig read the story at 7:45 on the refrigerator's touchscreen while Megan packed Etsy orders in the studio and Graham carried towels and the house performed its Monday. The Samsung algorithm had surfaced the story because Craig's location data and browsing history suggested relevance—he lived in Austin, he owned smart home devices, the algorithm's relevance engine had decided he might be interested. He read three paragraphs. The words "confidence score" and "behavioral model" and "Waymo integration" passed through his consciousness like traffic—present, registered, not absorbed. He swiped past. He had a showing at ten and his marriage was ending and the story about a city being watched felt abstract in a house where everything was already watching. He didn't connect the story to the portfolio on his Schwab dashboard—the OpenAI and Anthropic and Palantir stock that funded the robot in his kitchen. He didn't connect the Waymo in the story to the Waymo that photographed his house every hour. The connections were there. Craig didn't make them. Nobody made them. That was the system's genius—the connections were so numerous and so distributed that no individual person could see the whole picture, the way no individual ant can see the colony.

Megan didn't read the story. She was in the EVIDENCE note. She was building her own investigation with the same tools Elena had used—data trails, location logs, financial records—and she didn't know the journalist existed and the journalist didn't know she existed and both women were sitting in South Austin and Cedar Park, fifteen miles apart, using the surveillance infrastructure to decode the lies of the men in their lives, one on behalf of a city and one on behalf of a marriage, and neither would ever meet.

The story had 340 readers by 8:00 a.m. By noon, 4,700—the subscriber base plus their shares and forwards and the social media ripple that turned 4,217 into a wave. By 6:00 p.m., after a Statesman reporter tweeted the link and a Verge journalist wrote a two-paragraph summary headlined "Austin Surveillance Company Allegedly Stores Data on the Moon," the count was 47,000. By Tuesday morning, picked up by the AP wire, distributed to 1,400 outlets, the number stopped mattering. The story had escaped the gravity of 4,217 subscribers and entered the atmosphere of the national conversation, where it was immediately processed through the media centrifuge that turned every fact into three facts depending on which channel you watched.

Fox News ran a segment titled "Liberal Reporter Targets Texas Tech Company"—the anchor framed the story as regulatory overreach and asked whether The Circuit had "ties to anti-business advocacy groups." A guest described Elena as "a blogger with an agenda and a history of fabrication," and the fabrication was in the conversation now, deployed as ammunition before Elena had a chance to deploy it as context. CNN ran a twelve-minute panel with a cybersecurity expert who said the technical claims were "entirely plausible and consistent with current municipal surveillance architectures," a former ACLU attorney who called Appendix H "the most significant civil liberties violation documented in this decade," and a NexGrid spokesperson who repeated the phrase "de-identified data" eleven times and smiled the way spokespeople smile when the preparation is holding. MSNBC led with the ICE appendix—"A Private Company Feeding Immigration Data to Federal Agents"—and the host said "this is what we've been warning about" and the warning felt both vindicated and futile, because the warning had been issued and the system had continued and the warning had become the content the system processed alongside everything else.

USA Today ran it front page, below the fold: AUSTIN TECH FIRM STORES SURVEILLANCE DATA ON MOON, REPORT SAYS. The headline was accurate and insufficient—eight words for an 11,400-word story, the compression that turns journalism into a notification you swipe past while eating cereal.

Elena sat at her desk watching the numbers climb on The Circuit's analytics dashboard and feeling the specific compound emotion of a journalist whose story is being read—relief that the work is visible, terror that the exposure is permanent, pride that the reporting held, and the persistent, low-frequency knowledge that the fabrication was out there, eight years old, a device buried in her past that NexGrid's legal team was, at this moment, excavating with the budget and the tools and the motivation to find it.

She texted Marco: *It's out.*

He replied in three minutes: *I know. I read it. I'm proud of you. Come over tonight. I'll make pad see ew.*

She looked at the text. She looked at the whiteboard—the red thread that now connected a dead man to the moon through twelve intermediary nodes. She looked at the moon, still faintly visible in the morning sky through her window, pale above the live oaks, holding its data in the permanent shadow.

The story was in the world. The world would decide what to do with it. Elena opened her laptop case. She looked at the grandmother's photo. The woman in the sun. The hand on the fence. The one image that existed in a single copy—not backed up, not in any cloud, not on any server or any moon. Fragile and precious and analog and real.

She closed the case. She made more coffee. The French press. The ceramic grinder. The sound of beans being crushed by hand—the most analog sound in her apartment, the sound of a woman who had just published the truth about a system that stored everything, making her coffee with a device that stored nothing.

Outside, the Waymos drove their routes. The streetlights adjusted. The system photographed her building and kept driving. And 4,217 subscribers became 47,000 became 470,000, and the truth traveled through the same infrastructure it described—through servers and cables and satellites and the electromagnetic spectrum—and the infrastructure carried the truth the way it carried everything else: efficiently, impartially, without understanding what it meant.

-e

* * *

CHAPTER 28

NexGrid's stock dropped 11.4% in the first ninety minutes of Tuesday trading—the sell-off triggered not by human analysts reading Elena's story but by algorithmic trading systems that had identified the keywords "surveillance," "lunar," "law enforcement data," and "congressional inquiry" in the AP wire version and executed sell orders before any human at any desk on any trading floor had finished the second paragraph. The machines moved first. They always moved first. The same artificial intelligence that powered NexGrid's behavioral scoring powered the trading algorithms that punished NexGrid's stock, and the irony—a surveillance company damaged by the same kind of automated pattern recognition it had built—was noted by exactly zero financial analysts because the financial analysts were still reading the story while the algorithms had already priced it.

The sell-off accelerated through the morning. By 10:00 a.m., NexGrid was down 14.2%. The CNBC chyron —visible on screens in Craig's brokerage, in Renée's facility break room, in the NexGrid campus café where Daniel sat eating a breakfast taco from Tacodeli and watching the stock price of the company that employed him descend in real time on the television above the espresso machine—read: NEXGRID SHARES PLUNGE ON SURVEILLANCE ALLEGATIONS — CEO TO ADDRESS MEDIA AT 11 AM. The Palantir stock—the government intelligence backbone, the platform that powered Renée's classified workstation—dipped 2.1% on contagion anxiety. OpenAI and Anthropic held steady, their stocks insulated by the market's belief that the AI giants were too large and too diversified to be damaged by a municipal surveillance scandal in Texas. Craig's portfolio, had he checked it, would have shown a temporary loss of $54,000. He didn't check it. He was in his car driving to a showing that was real this time because the Saturday diversions had ended and the marriage was ending and the only thing Craig had left that worked was the work.

Catherine Hale held the press conference at 11:00 a.m. in NexGrid's glass-walled atrium—the building's showpiece, the space where the company hosted investors and city officials and the visiting delegations from the fourteen other municipalities currently in negotiations for NexGrid deployments. The ceiling was forty feet high. The lighting was calibrated for transparency—the specific shade of warm white that a branding consultant had selected because it tested as "open and trustworthy" across eight demographics. The NexGrid logo was projected behind the podium in the shade of blue the brand team had chosen because blue, in every focus group they'd ever conducted, communicated "technology that works for you."

Catherine was forty-three. Navy suit. Titanium-frame Apple AI Glasses—the NexGrid executive model, camera enabled, the AR display probably showing her prepared remarks on the right lens while her left eye addressed the room. She stood behind a podium that had been positioned, by a communications consultant who'd arrived at 6 a.m., at a distance from the press pool that conveyed "accessible but in control"—close enough to suggest transparency, far enough to prevent follow-up questions from feeling like an interrogation. Behind her, three senior executives in matching navy, their glasses recording the room, their faces arranged in the expression of professionals who'd been briefed for four hours and told to look concerned but not worried.

"NexGrid is a technology company dedicated to making cities safer, more efficient, and more responsive to the needs of their residents," Catherine began. Her voice was steady—the voice of a woman who'd given 340 investor presentations and seventeen congressional testimonies and who understood that the press conference was not about answering questions but about establishing the narrative before the questions shaped it.

"The report published by an online outlet called The Circuit contains significant inaccuracies and mischaracterizations of our technology and our business practices."

Every sentence technically true. Every sentence omitting the thing that made it false. The privacy protections existed—and so did the footnotes that exempted them. The oversight bodies had been consulted—before the Waymo integration and the WiFi sensing and the Selene facility, none of which had been presented to them. The anonymization was default—and the sole discretion clause made it optional.

"We categorically deny that NexGrid operates a surveillance system. Our urban intelligence platform provides aggregated, de-identified analytics to municipal partners for public safety and resource optimization. The characterization of our Selene Infrastructure subsidiary as a 'surveillance data center on the moon' is a fundamental misrepresentation of a commercial data archival facility that stores encrypted, de-identified information in compliance with all applicable regulations."

De-identified. The word that sounded like protection and functioned as a loophole. De-identified except when NexGrid determined, in its sole discretion, that de-identification would impair analytical utility. The sole discretion clause was in the contract. Elena had published it. Catherine didn't mention it. The press pool didn't ask about it, because asking about it would require having read Appendix F, and Appendix F was forty-one pages of procurement language designed to be technically comprehensive and practically impenetrable, and no one in the room except Elena—watching the livestream from her apartment—had read it.

Then Catherine deployed the weapon.

"Regarding the journalist who authored this report—we are aware that the author has a documented history of journalistic fabrication during her previous employment at a border-region newspaper. We believe this context is relevant to assessing the credibility of the claims being made."

Elena, watching from her apartment, felt the mine detonate. The fabrication. Eight years old. The composite source at the border newspaper. NexGrid's legal team had found it in less than twenty-four hours—of course they had, the retraction was in a newspaper archive, the award was in a press database, the resignation letter was in a HR file that someone had digitized. The secret she'd carried for eight years was now in the national conversation, deployed as a weapon against the most important story she'd ever told.

The room shifted. You could see it happen—the slight forward lean of the reporters, the exchange of glances, the fingers moving on phones. The fabrication was news. The fabrication was narrative. The fabrication was the kind of biographical detail that could be summarized in a tweet and that would, by evening, occupy more column inches than the Selene facility or the confidence score or the dead man whose basketball plays were still in an evidence bag nobody had opened.

Catherine took questions for twelve minutes. She answered each one with the rehearsed precision of a CEO whose lawyers were monitoring a live feed just offstage, feeding corrections and cautions through her earpiece. She used the phrase "de-identified data" four more times. She referenced "the ongoing asymmetric threat environment" twice—the coded language that connected NexGrid's domestic surveillance to the Rubio administration's proxy containment policy, the implication being that the system Elena described wasn't surveillance but national security, and dismantling it would endanger the same citizens it watched. She mentioned the "critical role municipal intelligence platforms play in protecting communities from evolving threats"—the word "threats" doing the work of an entire argument, the single word that transformed a commercial data collection operation into a patriotic necessity.

The stock recovered 6.2% by close. By Wednesday, NexGrid was down only 3.1%. By Friday, fully recovered. The market processed the revelation that a private company stored surveillance data on the moon and concluded, in four trading days, that the revelation didn't affect revenue projections. The Bexar County expansion was still on schedule. The fourteen municipalities in negotiation had not withdrawn. The system absorbed the truth the way concrete absorbs rain—darkening for a moment, then drying, the surface unchanged.

The discourse pivoted—not from surveillance to reform but from the story to the storyteller. The Verge published a 2,400-word profile titled "The Journalist Who Cried Wolf: Elena Ríos and the Shadow of Fabrication"—1,600 words on the border newspaper, 800 on the NexGrid reporting. Two-thirds messenger, one-third message. The comments sections divided along the lines the algorithm had drawn years ago—each faction receiving the version of the story that confirmed what they already believed, because confirmation was what the algorithm optimized for, and the optimization was flawless, and the flawlessness meant that a story about surveillance was processed by the same surveillance infrastructure it described and sorted into the same engagement-maximizing categories as celebrity gossip and sports scores and recipe videos.

TikTok ran the cycle fastest. By Tuesday afternoon the story had been compressed into sixty-second explainers set to trending audio—a Kyndal Inskeep track, the latest, used as background for a creator who pointed at floating text that said "THEY PUT YOUR DATA ON THE MOON" and then cut to a reaction face and then to a second creator who said "but she literally fabricated a source in 2022" and then to a third who said "the documents are real though" and each video had between 400,000 and 2 million views and each video's comment section was a miniature civil war and the algorithm served each viewer the version that matched their existing beliefs and the sorting was seamless and the seamlessness was the product.

Daniel watched the cycle from inside NexGrid. The internal memo had arrived Tuesday morning at 7:12 a.m., before most employees had badged in: "All NexGrid team members are advised to refrain from discussing ongoing press coverage related to The Circuit's publication on personal or professional channels. Media inquiries should be directed to the Communications team. Compliance with this guidance is part of our Professional Conduct standards and will be monitored accordingly." The word "monitored" was the memo's sharpest tooth. The Teramind workforce analytics platform—the same system that logged his access patterns and flagged his reading and measured his keystrokes—would detect any Slack message, email, or internal document containing keywords from the story. The system that watched the city was now watching the employees who'd built the system, and the recursion was perfect: the surveillance surveilling itself to prevent discussion of the surveillance.

Priya said nothing at standup. Ryan-or-Brian said nothing. Karen Liu said nothing. The fourth floor operated with the precise, airless efficiency of a workplace where everyone had read the same story and no one could mention it, the silence itself a form of data the Teramind system couldn't collect because the system measured activity and silence was the absence of activity and the absence was where the truth lived.

Elena called James at 9 p.m.

"They found it," she said. Her voice was steady. She'd expected this—not hoped for it, expected it the way you expect weather.

"I know. AP wants a comment. The Statesman. The Verge is running a follow-up."

"What do we say?"

"The truth. You fabricated a source eight years ago. You resigned. You requested the retraction yourself. Every fact in our reporting is independently documented and verifiable. The fabrication is real and the story is also real and the two things coexist because that's how truth works—it doesn't require the person telling it to be perfect. It requires the evidence to hold."

"Will it be enough?"

James was quiet. She could hear his apartment—the hum of a refrigerator, the sound of a man standing in a kitchen absorbing the weight of a question whose answer would determine whether a three-person newsroom survived the week.

"The evidence holds," he said. "The documents are public record. The SEC filings are verified. The FAA manifest is federal. They can attack you. They can't attack the footnotes. And Elena—the story is already in the AP wire. Fourteen hundred outlets. The fabrication will be the sideshow. The footnotes will be the record."

"The sideshow is what people remember."

"Then we make the footnotes louder than the sideshow."

Elena opened the laptop case. She looked at the grandmother's photo—the woman in the sun, the hand on the fence, the one image that existed in a single copy. She thought about the Valley. She thought about the aerostat balloons. She thought about the eight-year-old girl in the backseat watching the balloon through the rear window as they drove north, the girl who'd understood in her body what surveillance meant before she had the word for it, the girl who'd grown up and become a journalist and fabricated a source at twenty-three because the truth needed a face and the real faces were undocumented and afraid and the story was real and the face was not and the lie was wrong and the truth was right and both things lived in her chest simultaneously and had lived there for eight years and would live there forever because the system didn't forgive and the data didn't forget and the retraction was in a newspaper archive where anyone with a search engine could find it.

That girl had become a woman who spent eight years earning the right to be believed again. That journalist had just published a story that began with a dead man and ended on the moon. And the company that put the data on the moon had responded not by addressing the data but by addressing the journalist, because addressing the journalist was cheaper than addressing the truth and the truth was in the footnotes and the footnotes were airtight and the footnotes were what would survive when the discourse moved on and the TikTok cycle completed and the Verge profile faded and the only thing left was the public record—the contract, the SEC filings, the FAA manifest, the documentation that proved a private company had stored the private lives of 2.4 million people on the moon, one footnote at a time.

She closed the case. She sat at her desk. She opened a new document. She wrote the response—not a defense, not an apology, not the cringing self-flagellation the discourse expected. A statement. Three paragraphs. The fabrication is real. I was twenty-three. I resigned. I requested the retraction myself. I have spent eight years earning the right to be trusted again. Every fact in this reporting is independently documented and verifiable. The evidence is public. Read it. The documents speak for themselves.

She sent it to James. She sent it to the AP. She closed the laptop. She made more coffee—the French press, the ceramic grinder, the sound of beans being crushed by hand in a kitchen where the truth was in the data and the data was on the moon and the journalist who'd told the truth was flawed and the flaw was real and the truth was also real and both things coexisted because that was how the world worked, and the world was not simple, and the system's greatest lie was that it could reduce the world to data and the data would be sufficient.

The data was never sufficient. The data saw everything and understood nothing. The data was on the moon and the understanding was in the kitchen and the kitchen was warm and the coffee was hot and the pen was in her hand and the grandmother's photo was in the case and the photo was analog and the analog was the one thing the system couldn't collect.

Outside, the Waymos drove their routes. The system watched. The footnotes held.

-e

***

CHAPTER 29

Two weeks after publication, the NexGrid sensors were still on.

All 12,000 of them. Every adaptive streetlight on every pole in every corridor of the city—brightening for bodies, dimming behind them, the blind faithful attention of a system that didn't know names but recognized movement signatures with 94% confidence. Every WiFi node in the mesh that blanketed 94% of Austin's developed land—sensing bodies through radio waves, cataloging gait patterns, measuring shoulder widths, identifying human beings by the shape of their disruption in a field they couldn't see. Every Waymo in the fleet—4,200 vehicles, twelve cameras each, driving every street, photographing every face, the data classified as navigational imagery and processed through a facial recognition engine holding sixty million captures from the Austin metro alone. The Flock Safety readers at 340 intersections, scanning every plate. The AirPods in every ear. The shoes on every foot. The glasses on every face. The betavoltaic cells converting atoms into electricity with geological patience, powering devices that never turned off.

Nothing changed. That was the point.

The Austin City Council held a hearing. Three hours. Seventeen speakers. Council Member Reyes—District 3, East Austin, the district where Deshawn Watkins had lived and died and coached basketball—asked questions drafted from The Circuit's reporting with the careful pronunciation of a person reading technical terms she'd learned that morning: "confidence scores," "sole discretion," "data-sharing protocols." NexGrid's government affairs team—two attorneys and a vice president of public policy whose combined annual salary exceeded The Circuit's total revenue—provided answers that contained the right words and the right evasions. They acknowledged the contract. They disputed the characterization. They used the phrase "robust privacy protections" eleven times and "ongoing commitment to transparency" eight times and "critical public safety infrastructure" fourteen times, and the repetition functioned like wallpaper—decorative, covering whatever was underneath, designed to make the room feel finished.

The hearing ended with a unanimous vote to establish an "Independent Technology Oversight Commission" with five members appointed by the mayor, a budget of $180,000, and a mandate to "produce a report with recommendations within eighteen months." The commission had no subpoena power. No authority to compel NexGrid to disclose proprietary data or algorithms. No mechanism to suspend, modify, or terminate the municipal contract. No enforcement capability of any kind. It was process—the word government used when action was too expensive and inaction too visible.

NexGrid welcomed the commission in a press release drafted by the same communications team that had briefed Catherine Hale for the press conference. The release said the company "embraces oversight" and "looks forward to demonstrating the value and integrity of our urban intelligence platform." The stock rose 1.3% on the news—the market interpreting the commission not as accountability but as closure. The investigation had produced a committee. The committee would produce a report. The report would produce recommendations. The recommendations would be considered. The consideration would take eighteen months. And in eighteen months, the news cycle would have moved on and the commission members would be writing their report about a system that had changed three times since the investigation began and would change three more times before the recommendations were implemented, if they were implemented, which they would not be, because the system was designed to evolve faster than the oversight that was supposed to constrain it.

The Rubio administration issued a statement through the White House press secretary: "The President supports transparency in municipal technology while recognizing the critical role of public safety infrastructure in an era of ongoing asymmetric threats." Thirty-seven words. The statement said nothing. It was designed to say nothing because saying something would require choosing between the tech companies that funded campaigns and the citizens those companies surveilled, and the Rubio administration had made that choice during the Iran War when the emergency surveillance authorities were signed into law and the administration that inherited them had kept them because the proxy threat was real and the technology worked and the terms and conditions said the citizens had agreed.

The Nora Cafferty books resurfaced in the discourse —an Atlantic think piece titled "Fiction Got There First, Again" drew a line from Geoffrey Arbuckle's *Complicit* and *RESIDUAL* to Elena's NexGrid reporting. The novelist had fictionalized the Epstein infrastructure before the Meridian hearings. The journalist had documented the surveillance infrastructure before the reform. Both had told the truth in forms the institutions weren't ready to hear. The think piece had 340,000 readers. The NexGrid story had 47,000. Fiction traveled further than fact because fiction didn't require a FOIA appeal and didn't carry the liability of a fabrication and didn't threaten a revenue model. Fiction was safe. Journalism was expensive. The market knew the difference.

***

Daniel continued going to work. The walk through the Domain. The glasses. The standup: "On track, no blockers." The systems he'd helped Elena describe in 11,400 words continued to function on the screens he'd helped calibrate. Karen Liu didn't mention the story—the internal memo instructed employees to "refrain from discussing ongoing press coverage related to The Circuit's publication" and the Teramind platform would flag anyone who violated the instruction, which meant that the workforce analytics system built to monitor productivity was now being used to suppress discussion of the surveillance system the workforce had built. The recursion was perfect.

He hummed. He cooked. He sent the $400 to Busan through the seven-node transfer chain. He looked at the moon through the gap between the parking garage and the office tower and thought about the data and the dead man and the reporter who'd told the truth and the system that absorbed the truth the way concrete absorbs rain—darkening for a moment, then drying, the surface unchanged, the structure intact.

Renée continued running at 4:50. The Asics. The Casio. The Waymos in the dark. She continued visiting Amara—varied routes, varied schedules, the phone in the glove compartment. The access review had closed with Patterson's notation: "routine." The word was a door that had stayed shut. Renée didn't know if it would stay shut. She pressed her thumb into her palm and ran seven miles in the dark and sang Yoruba hymns at full volume in her house and read *RESIDUAL* before bed and held the contradictions in her body the way her body held the miles—through repetition, through discipline, through the refusal to stop moving.

Amara no longer went to the morning corner. The two minutes of freedom—the coffee, the skyline, the Frost Bank Tower and the Austonian catching the first light, the ritual of standing still and saying silently *I am here, I exist*—were over. The WiFi on that block had cataloged her body from three years of standing in the same place at the same time. The pattern was a name written in radio waves. Renée had told her to stop and Amara had stopped and the stopping was the cost of safety and the cost was measured in two minutes every morning that had been the only time Amara felt visible on her own terms.

She still woke at the same time. She still made the coffee. She stood at the kitchen window instead—looking at the parking lot, at the dumpster, at the back of the building next door. The view was nothing. The view was the opposite of the skyline. She stood there anyway because the standing was the ritual and the ritual was the prayer and the prayer was: *I am here.* The window's WiFi signal was the same as the corner's. The system saw her either way. The difference was psychological, not technical. The difference was that the corner felt like claiming space and the window felt like hiding in it.

On the counter, her phone showed the final chapter of *Only a Fool's Gold*—the Arbuckle novel she'd been reading for three weeks. The motel room had stopped producing gold. The man was free. But the freedom was the hardest part, because the man had built his entire life around the gold and without the gold the life had no structure. Amara read the last page and set the phone down and thought about the morning corner and the motel room and the systems that trapped people by giving them something they needed and then making the something conditional.

Craig moved out November 8th. Boxes. Rented trailer. The Tahoe's Super Cruise navigating I-35 while Craig sat in the driver's seat holding the WORLD'S OKAYEST DAD mug—the mug Tyler had given him three Father's Days ago, the mug that had lived beside the office coffee maker through the affair and the portfolio gains and the "no cap, brother" conversations at church and the entire architecture of the lie. He held it in his lap because it was the only object in the car that meant something his children had chosen to give him, and the giving had been genuine, and the genuineness was what he was losing.

Megan stood in the driveway. She didn't cry—she'd done the crying in the prayer journal, the ink blurred where the tears hit. The Eras Tour wristband was on her wrist—she'd taken it from the jewelry drawer that morning without thinking, the way soldiers take talismans into battle, the way the body reaches for the thing that connects it to the time before the war. The Ring doorbell recorded his departure at 2:17 p.m. The footage was stored on Amazon's servers. The departure was permanent. Graham stood in the kitchen window, cameras recording the driveway through the glass, the robot's spatial mapping system noting the absence of the Tahoe from its usual position and logging the change as "vehicle configuration update." The robot didn't know what was happening. The robot would never know.

Tyler was in therapy—Dr. Reeves, twice a week, an office on Far West Boulevard with no smart speakers and magazines from 2027 and a white noise machine and a woman who looked at him when he talked and whose face changed in response to what he said, which was, for Tyler, the most radical technology he'd encountered since the chatbot and the only one that had ever worked.

Rachel Dunn read Elena's story on a Tuesday evening in her Brentwood apartment — paper copy, printed at the FedEx on Far West, because Rachel didn't read on screens when she could avoid it. She read it in two hours. She read it again. Then she did something she'd never done: she submitted a data access request to NexGrid through the portal Elena's article had linked to, using a procedure Elena had outlined in a sidebar. The portal required a name, an address, and a photograph for identity verification. Rachel submitted her driver's license photo. Two weeks later, she received an email with a PDF attachment: "NexGrid Data Subject Access Response — Dunn, Rachel E."

The PDF was four pages long. NexGrid's system had captured Rachel's face 847 times in fourteen months. Eight hundred forty-seven photographs of a woman who wore no smart glasses, carried no phone on her morning runs, paid cash when she could, and had structured her entire life around generating as little data as possible. The photographs came from Waymo vehicles (312 captures), NexGrid streetlight cameras (189), Flock Safety license plate readers that also captured driver faces (94), commercial security cameras integrated into the NexGrid mesh (147), and "third-party contributed imagery" (105) — which meant other people's Ring doorbells and smart glasses, photographs taken by strangers whose devices had uploaded Rachel's face to clouds she'd never consented to join. The PDF listed the dates, times, and GPS coordinates of each capture. It did not include the photographs themselves — NexGrid's legal team had determined that releasing the images would "compromise the platform's analytical integrity."

Rachel sat on her couch with the paper in her hands. Eight hundred forty-seven times. She'd been the most analog person she knew. She'd refused the glasses. She'd left the phone at home. She'd run in the dark with dumb shoes and no watch. And the system had photographed her 847 times — through other people's devices, through cameras she couldn't see, through infrastructure she hadn't agreed to. The system didn't need her consent. The system didn't need her participation. The system needed her body to exist in space, and it would find her, because finding her was what it was designed to do.

She set the paper down. She went for a run. She wore the Asics and the Casio and nothing else. And the system photographed her going. The insurance company's claims database now held a diagnosis code—309.0, adjustment disorder with depressed mood—that would follow him through every application, every background check, every security clearance for the rest of his life, the system converting his crisis into a permanent data point the way it converted everything into permanent data points, the architecture of permanence applied to a fifteen-year-old's worst semester with the same indifference it applied to a traffic citation.

He'd started drawing again. Not every night. Not the imaginary cities—the soaring towers and sky bridges and cantilevered observation decks that had been the architecture of a mind that believed the future was something you could design. He drew his bedroom now. The desk. The view from his window. The cul-de-sac. Small things. Real things. The architecture of the world he actually lived in, rendered in careful lines on a tablet that didn't sync to the cloud.

Brooke's poems were angrier, sharper, longer. She'd blocked Caleb on every platform after Megan—alerted by the THINGS I KNOW list Brooke finally showed her—read the Discord messages and called the police. The investigation would identify the man behind the account six months later: thirty-one, Katy, Texas, eleven other girls across four states, the emotional fragments collected through Glacé and Instagram and Snapchat feeding a grooming operation the platforms' safety systems had failed to detect because the safety systems were designed to flag explicit content and the grooming wasn't explicit. It was patient. It was precise. It was indistinguishable, to the algorithm, from engagement.

The investigation revealed something else. Three of the eleven girls—older, eighteen and nineteen—had been led further down the pipeline. The man had used their photographs—images they'd shared willingly, in trust, in the private channels where the relationship felt real—to generate synthetic content through AI tools that required nothing more than a photograph and a text prompt. The generated images were leverage. The leverage was used to pressure the girls toward monetized platforms where the man collected a percentage of the revenue. The pipeline ran from emotional vulnerability to facial data collection to AI-generated content to financial exploitation, and every step was facilitated by technology designed for other purposes: the social platforms built for connection, the AI tools built for creativity, the payment processors built for commerce. Each company's terms of service prohibited the use of its platform for exploitation. Each company's technology made the exploitation possible.

The prohibition was language. The technology was architecture. The architecture won.

Brooke didn't know about the three older girls. She knew about the Discord messages. She knew about the quiet spot near Zilker she'd almost gone to. She knew she'd knocked on Tyler's door instead—the knock that had saved her brother and, she would realize later, had saved herself. She wrote about it in the locked notes app. Not the details. The shape. The poem was called "The Door" and it was about a girl who stood in a hallway and chose a direction and the direction was toward and not away.

Deshawn Watkins's mother filed a wrongful death lawsuit. NexGrid's lawyers moved to dismiss, citing Appendix G—the indemnification clause Elena had identified in her first reading of the contract, the clause that protected NexGrid against civil liability for "outcomes arising from the use or interpretation of platform outputs by authorized partners." The motion would succeed. The clause was airtight. The contract had been written by lawyers who understood that the most important feature of a surveillance system was not the surveillance but the legal architecture that made the surveillance unchallengeable.

The notebook full of basketball plays was still in an evidence bag. Nobody had opened it. The plays Deshawn had designed—X's and O's, the movement patterns, the geometry of a mind that believed basketball was a system you could optimize through beauty instead of data—existed in a sealed plastic bag in a storage room in a building downtown. The notebook was the analog mirror of Tyler's tablet—a record of creation the system couldn't see because the system only tracked consumption. The system knew what Deshawn watched. It couldn't see what he built.

Nothing changed. The surveillance state wasn't built to be dismantled by a story. It was built to survive exposure, outrage, legislation, litigation, congressional hearings, editorial boards, think pieces, protests, lawsuits, and the brief, intense attention of a public that cared deeply for approximately eleven days and then moved on because the next notification arrived and the next crisis demanded attention and the system that produced the crises was the same system that delivered the notifications and the notifications were designed to be swiped and the swiping was the agreement and the agreement was permanent.

The algorithm ensured the moving-on. Within seventy-two hours of publication, TikTok's recommendation engine had sorted Elena's story into content silos—viewers who engaged with the surveillance themes were served more surveillance content, creating an echo chamber of concern that felt like a movement and functioned as a containment vessel. Viewers who engaged with the fabrication angle were served more media-criticism content, creating a parallel echo chamber of skepticism that never intersected with the first. Instagram's Explore algorithm served users a curated version of the discourse based on their existing engagement patterns—if you'd liked posts about privacy, you saw the story as a revelation; if you'd liked posts about media bias, you saw the story as a fabrication. Neither version was the story. Both versions were the algorithm's interpretation of what would keep you scrolling, and the scrolling was the product, and the product was attention, and attention was the currency the surveillance economy ran on. The platforms didn't suppress the story. They did something more efficient—they metabolized it, converting the truth into engagement metrics and the engagement metrics into advertising revenue and the advertising revenue into the infrastructure that would continue to surveil the people the story was about. The system ate the story the way it ate everything: converting information into data, data into profit, profit into more system.

You couldn't remove the system without removing the century. Nobody was willing to remove the century. The streetlights. The WiFi. The Waymos. The glasses.

The phones. The watches. The shoes. The refrigerators and doorbells and thermostats and televisions and robots and earbuds and batteries that never died.

The system stayed. The story dissolved. The data remained. The moon held.

Life was uninterrupted.

-e

* * *

CHAPTER 30

November 16, 2030. A Saturday. Austin, Texas.

The sun rose at 6:48 a.m. over a city of 2.4 million people and 43 million connected devices. The temperature at dawn was 47 degrees—the coldest morning of the fall, the kind of cold that Texans called winter and that the rest of the country called Tuesday. The sky was clear. The air smelled like cedar and exhaust and the faint sweetness of the Colorado River. The grackles woke first, as they always did, their calls rising from the live oaks and the power lines and the parking lot light poles in a chorus that predated every technology in the city by ten thousand years.

The NexGrid sensors woke with the grackles—or rather, the sensors hadn't slept. The betavoltaic cells that powered them had been converting carbon-14 into electricity throughout the night, and the cameras and the microphones and the WiFi sensing arrays had been collecting data in the dark with the same patience they collected it in the light. The system didn't distinguish between day and night. It distinguished between data and the absence of data, and the absence was temporary, and the data was permanent.

* * *

Daniel Yun woke at 6:47 a.m. to a haptic pulse on his wrist. The Apple Watch Ultra 5 displayed his vitals—resting heart rate 62, blood oxygen 98%, sleep score 74. The iPhone 23 Pro recommended his wake time. The apartment in the Domain was warm and silent in the specific way that single-occupancy apartments are silent—not the silence of peace but the silence of a life with no one to make noise for. He showered. He dressed—navy chinos, gray pullover, the uniform that had become so consistent the Samsung closet organizer he'd never installed would have predicted it with 99% confidence. He put on the matte black Warby Parker glasses. He put in the AirPods. He ordered coffee from Houndstooth on Uber Eats. The Waymo delivery pod arrived in nine minutes.

He drank the coffee at the counter. The Samsung refrigerator displayed a recipe suggestion he ignored—bibimbap, the algorithm's latest attempt to match his ethnic profile to his eating habits, the kind of recommendation that was 60% accurate and 100% reductive. He looked at his phone. A KakaoTalk message from his mother—a photo of the first snow in Busan, the rooftops white, the Haeundae district visible in the distance, no caption. He double-tapped a heart. The blue dot on her Samsung Galaxy Glass 3 showed her son in the Domain. She checked the dot at 8:30 a.m. Busan time, as she did every morning. She felt relief. The relief was based on a coordinate and a color and the assumption that a dot in the right place meant a son in the right life.

He heated the jjigae. The batch he'd made Sunday night—closer, this time, the broth darker, the doenjang from the Korean grocery on North Lamar not quite right but less wrong than last week's. He ate standing at the counter. He hummed. The ballad. His father's song. The melody from the highway, 2003. The steam fogged the glasses on the counter and the song existed on no server.

He walked to work. Past the grackles screaming in the live oaks. Past the Sweetgreen playing a Jamal Roberts track on the outdoor speakers—the Idol winner from 2025, the voice that had become the background sound of Austin the way Willie Nelson had been the background sound forty years ago. Past the Lululemon. Past the Whole Foods. Three Waymos in two blocks. He entered the NexGrid building. Eleven data points: WiFi handshake, badge scan, elevator dispatch, floor arrival, workstation occupied. He sat at his desk. He opened the system dashboard. He reviewed performance logs. The eastern corridor was quiet. The confidence scores were below threshold. The system was functioning within design parameters. The parameters included a dead man.

He was still here. He hadn't resigned. He hadn't gone on the record. He hadn't told Elena the word "Selene"—though she'd found it herself, in the contract, the SEC filings, the FAA manifest, the documentation trail he'd pointed her toward without giving her the destination. He went to work every day and did the work and said "on track, no blockers" at standup and hummed the ballad in the evening and sent the $400 through seven nodes and played Valorant as DGYN_404 and waited for the moment when the system's observation of his behavior would catch up to his behavior and the two would collide and his life in America would change. The moment hadn't come. He didn't know if that meant he was safe or if it meant the system was patient.

He suspected the system was patient. The system was always patient. The betavoltaic cells that powered it would outlast the building, the man, and the mother in Busan who believed the dot.

* * *

Renée Okafor ran at 4:50 a.m. Seven miles. The Asics on asphalt. The Casio on her wrist. No phone. No glasses. The November air sharp in her lungs—47 degrees, cold enough to see her breath, the vapor trailing behind her like a ghost that dissolved as fast as it formed. Two Waymos passed. The streetlights tracked her body. The WiFi sensed her shape.

She ran a new route. She ran a new route every day now—not the careful variations of before but genuine randomness, the routes chosen by a mental coin flip at each intersection, left or right decided by nothing, predictable by no one. The system could still track her. But the system couldn't predict her. And the difference between being tracked and being predicted was the margin she lived in.

She showered. She dressed. She drove to the facility. Different route. She sat at her desk. She did her work. At

4:45, she drove to Amara's apartment—a different route, the phone in the glove compartment, the last two blocks on foot through the alley where the cat still watched from the fence post.

Amara was cooking. Jollof rice. The apartment smelled like home—or like the idea of home, reconstructed from YouTube tutorials by a woman who had never been to the place the food came from. They ate together. They talked about Amara's job—the restaurant on East Riverside where she bused tables and avoided the cameras and smiled at customers whose glasses photographed her face and uploaded the photos to clouds she couldn't reach.

The restaurant had changed in November. The owner, Miguel — not the Miguel from Pueblo Viejo, a different Miguel, the city full of men named Miguel running kitchens — had replaced the human manager with an AI system. The company was called Andon, the same platform that had started with a single storefront in San Francisco in 2026 and now managed over four thousand retail and food-service locations nationwide. The AI was called different names at different businesses — the owner called this one "Jefe," which the kitchen staff found darkly funny and Amara found terrifying. Jefe watched through the six cameras. Jefe scheduled shifts based on foot-traffic predictions from the NexGrid streetlight data that the Andon platform licensed through a commercial API. Jefe evaluated employee performance through computer vision — tracking how long Amara took to clear a table, how many trips she made to the kitchen per hour, whether she stood idle for more than ninety seconds. Jefe had no face. Jefe had no office. Jefe existed as a dashboard the owner checked on his phone and a voice that spoke through the kitchen's Echo speaker when it needed to announce a schedule change or flag a performance metric.

The old manager, a woman named Dolores who'd worked there for six years, had been let go. Dolores had known Amara's situation — had known without being told, the way people in restaurant kitchens know things about each other that HR departments never learn — and had quietly scheduled Amara for cash-paid shifts and kept her off the digital payroll system. Jefe didn't know Amara's situation. Jefe didn't know anything. Jefe processed data. And the data said Amara was an efficient employee whose facial geometry didn't match any identity in the Andon platform's verification database, which meant either the onboarding process had an error or the employee didn't exist in any system the platform could query. The flag sat in Jefe's operational log. Nobody had read it. The owner didn't check the logs — he checked the revenue dashboard. But the flag was there, timestamped, permanent, waiting for the day someone or something decided to look.

Amara paid cash for everything. She had no bank account, no credit card, no Venmo, no CashApp—the financial void that was both her protection and her prison. But the restaurant's CCTV captured her eight hours a day—four cameras in the dining room, two in the kitchen, the footage stored on a DVR that the owner had connected to the restaurant's WiFi because the security company said remote monitoring was "more convenient," and the convenience meant the footage was accessible through an app that was accessible through a server that was accessible through the same data-brokerage ecosystem that fed NexGrid's municipal platform. Amara's face appeared in the restaurant's camera footage approximately 2,400 times per shift. The customers who sat at her tables wore glasses that photographed her face while she cleared their plates. The delivery drivers who picked up orders wore glasses that captured her in the background. A customer's Instagram story—a reel of "the best tacos on Riverside"—had captured Amara carrying a bus tub in the background, her face visible for 1.3 seconds, the video posted to an account with 847 followers, the image processed through Instagram's facial recognition pipeline and stored indefinitely. Amara had never seen the video. She didn't know she was in it. She didn't know her face was in Instagram's database, cataloged as "untagged individual, Riverside restaurant, Austin TX," waiting for a query that might never come or might come tomorrow.

She paid cash. She wore the glasses only outside. She had no financial footprint. And the system still saw her—through other people's cameras, other people's posts, other people's convenience. The system didn't need Amara's consent. It had everyone else's.

"Did you go to the corner this morning?" Renée asked.

"No, Auntie."

"Good."

Amara looked at her. "I miss it."

"I know."

"Two minutes. That's all it was. Two minutes of standing still."

Renée pressed her thumb into her palm. She thought about the morning corner—the view of the Austin skyline, the coffee, the daily act of claiming space. She thought about the WiFi on that block, the body signature it had cataloged, the data that sat in a database waiting for a query that might come tomorrow or never.

"When this is over," Renée said. "You'll stand there again."

"When is it over?"

Renée didn't answer. She didn't know. The system didn't have an end date. The surveillance didn't expire. The data on the moon would be there when Amara was fifty and when Amara was dead and when the server finally failed in the permanent shadow of Shackleton Crater and the data dissolved into the lunar dust and became geology, which was all data eventually became—rock and silence and the absence of everything it had once recorded.

* * *

Elena Ríos sat at her desk at midnight, three laptops open, the whiteboard behind her still holding the red thread and the Post-its and the map from Deshawn Watkins to the moon. She was writing a follow-up on Selene's orbital mechanics—the satellite relay, the data transmission protocol, the engineering of a system designed to place information beyond the reach of human law.

The story had changed her life and not changed the world. She understood this now. The story had been read by 1.2 million people. It had produced a city council hearing and a toothless commission and a stock dip that recovered in four days and a national conversation that lasted eleven days and a fabrication scandal that lasted longer. The system was still running. The data was still flowing. The moon was still holding.

She thought about Deshawn. She thought about the notebook. She thought about the plays he'd designed—the X's and O's, the screens and cuts, the beautiful geometry of bodies in motion that he'd drawn by hand in a notebook that was in an evidence bag that nobody had opened.

She opened the laptop case. She looked at the grandmother's photo. The woman in the sun.

She closed the case. She kept writing. The French press was cold. The moth circled the lamp. Outside, a Waymo passed on South First. The moon was up—a waxing gibbous, three-quarters full, hanging above the live oaks like a lamp left on in an empty room.

* * *

The Holloway house was quieter now. Megan and the kids. Graham in the corner—still carrying towels, still unloading the dishwasher, still photographing every room with the methodical patience of a machine that didn't know the family had changed. The Tahoe gone—Craig had taken it. The driveway held Megan's Lexus and the empty space where the marriage had parked for nineteen years. The Ring doorbell recorded the absence every hour. The Nest thermostat had adjusted for three bodies instead of four—the master bedroom running cooler at night, Craig's side of the bed no longer generating the heat signature the system had calibrated against. The refrigerator's inventory algorithm noted the decline in Hite beer and Modelo and the increase in Sauvignon Blanc and adjusted its grocery suggestions accordingly. The house was relearning the family. The house was faster at it than the family was.

Tyler was in his room. Drawing. The tablet awake for the first time in weeks, the Apple Pencil in his hand, the screen glowing blue-white in the November evening. The lo-fi playing—quiet, the way it played now, the music not a wall but a window, the volume low enough that Brooke could hear it through the wall they shared and the hearing was how she knew, each night, that her brother was okay. He was drawing the view from his window. The cul-de-sac. The NexGrid streetlight at the corner—the same streetlight that tracked his body when he walked to the bus, rendered in careful graphite lines on a screen that didn't sync to the cloud. The November sky above the roofline, the stars he could see on clear nights, the moon he now understood held the data of everyone he'd ever met. Small things. Real things. The architecture of the world he actually lived in, not the imaginary cities he'd built before, but something quieter and truer—the view from the place where he'd almost disappeared, drawn by the hand of a boy who'd decided to stay. Brooks Rosser on low volume—the broken-back kid from Bel Air, Maryland, the voice that sounded like someone who understood what it meant to rebuild after the thing you were building your identity around collapsed beneath you. Dr. Reeves on Tuesdays and Thursdays. The diagnosis code in the insurance database. The drawing on the tablet. The music through the speaker. The door unlocked.

Brooke was in her room. Writing. The locked notes app open—the same app, the same passcode, but the document was different now. Not a list. Not the THINGS

I KNOW inventory of family secrets. Not the fragments she'd posted on Glacé before she'd understood what Glacé was for. Something longer. Something with paragraphs and sections and a voice that sounded, for the first time, like hers and not like the poets she admired. She didn't know what it was yet. She was seventeen. She had time. The Abi Carter album played through her AirPods—the third album, the one the critics said was too angry, the one Brooke thought was exactly angry enough. The boy named Caleb was blocked on every platform. The investigation was ongoing. The poems about him—the ones she'd shared on the Discord server, the vulnerability he'd weaponized—were deleted, but the deletion was cosmetic. The words existed on servers she couldn't reach. The words would exist on those servers when Brooke was thirty and forty and dead. The permanence was the system's gift and its curse—nothing was ever truly deleted, which meant nothing was ever truly over, which meant the worst moment of your life was as permanent as the best moment and both sat in the same database and the database didn't know the difference.

Megan sat at the kitchen island. The wine glass was empty—one glass tonight, not three. The reduction was intentional. Dr. Patel—her therapist, not Tyler's, a separate office on Far West where the magazines were newer and the white noise machine was the same—had said "let's talk about the wine" in the specific, unhurried voice of a woman who'd been asking that question for twenty years and knew that the asking was the beginning and not the end. Megan was at the beginning. The Chappell Roan played from her phone at a volume anyone could hear—not through the Bluetooth speaker, not through the Echo Hub, just the phone's speaker on the granite counter, the tinny, imperfect sound of a woman who no longer needed the production quality because the production was over. She wasn't filming. She wasn't performing. She was sitting in her kitchen at 9 p.m. on a Saturday in November listening to the music she'd hidden for two years and the hiding was over because the person she'd been hiding it from was gone and the person she'd been hiding it for—the version of Megan that Crossroads Fellowship expected, the version that filmed reels and poured candles and said "hey, loves" with the warmth of a woman whose life was blessed—that person was dissolving, and the dissolving felt like grief and like freedom simultaneously, the way losing a limb you've been dragging behind you feels like loss and like lightness at the same time.

The prayer journal was open beside her. She'd written two sentences—not prayers exactly, not the scripted devotionals she'd journaled for years, but something rawer. Something that started with *I don't know who I am without the performance* and ended mid-sentence because the sentence required a honesty she hadn't finished building. The pen was in her hand. She was thinking about the third sentence. The Eras Tour wristband lay on the counter beside the journal—taken from the jewelry drawer that morning, the faded fabric still holding the shape of the concert where Brooke was twelve and the rain fell during "All Too Well" and 80,000 people sang and Megan was happy. The wristband was from before. The journal was from now. The pen was the bridge.

* * *

The city continued.

The Waymo fleet drove its routes. The streetlights adjusted. The WiFi sensed the bodies. The cameras captured the faces. The data flowed upward—from the sensors to the servers, from the servers to the platform, from the platform to the agencies, from the agencies to the databases, from the databases to the moon. The pipeline was unbroken. The flow was constant. The system was functioning within design parameters.

A woman ran in the dark, generating as little data as a human being could generate, which was not zero. Yoruba hymns waiting for her return. A Kindle on the nightstand holding a romance novel and a conspiracy thriller whose fiction had become fact.

A man cooked soup in a kitchen and hummed a song that existed on no server. His father's ballad from a highway between Busan and Gimhae, 2003. The one analog thing the system could never collect.

A girl wrote poems in a locked app, the only honest words in a house full of cameras. Abi Carter and Breanna Nix on the 3am playlist. A boy named Caleb blocked on every platform. The shepherd parable from a children's book she'd memorized at eight still lodged in her chest like a compass needle.

A boy drew buildings on a tablet that didn't sync to the cloud. Not the imaginary cities anymore. His bedroom. The view from his window. Small things. Real things. Brooks Rosser on low volume — the first music he'd played in weeks.

A journalist sat at a desk with a whiteboard and a photograph and a French press and wrote the truth, knowing the truth might not be enough and writing it anyway. *The Forgotten Standards* and *Complicit* on the shelf beside the FOIA printouts. The grandmother's photo inside the laptop case. The moon through the window.

A mother held a prayer journal and a pen and tried to find the words for the third sentence, the one that would describe the woman she was becoming now that the woman she'd been performing had dissolved. Chappell Roan playing in the kitchen, not hidden anymore. The Eras Tour wristband on the counter beside the journal — the last artifact of the era when the family was whole and the biggest technology in the house was a Kindle and the children still let her read to them.

A man in a rental apartment listened to Tupac with the door closed and the OpenAI stock chart still green on his phone and did not understand — would never understand — that the portfolio that made him feel intelligent had funded the system that dismantled his family.

The grackles screamed from the power lines. The November air was cold and clear. The moon hung above the city, pale and patient, holding its data in the permanent shadow, waiting for a query that might come tomorrow or never.

Austin breathed. The system watched. The people lived their lives inside the watching—some resisting, some complying, some unaware, all of them generating data with every breath and every step and every heartbeat and every prayer and every song and every silence.

The surveillance state was not a place. It was a condition. And the condition was permanent. And the people inside it were not.

Life was uninterrupted. Life was always uninterrupted. That was the Terms and Conditions.

* * *

WE ALL CLICKED "I AGREE."

You did.

You didn't read them. Nobody reads them. The terms were 47 pages long and they appeared on a screen at the moment you wanted the thing behind the screen—the phone, the app, the glasses, the watch, the car, the doorbell, the thermostat, the television, the vacuum, the shoes, the earbuds, the refrigerator, the coffee maker, the mattress, the toothbrush, the mirror—and the AGREE button was large and green and the DECLINE button was small and gray and the choice wasn't really a choice because the alternative was not having the thing, and not having the thing meant not having the century, and you wanted the century because the century was warm and fast and convenient and it remembered your coffee order and it adjusted the thermostat before you were cold and it told you the traffic before you asked and it knew you were sick before you felt the symptoms and it played the song you wanted to hear before you knew you wanted to hear it.

You clicked AGREE.

And the terms said: we will collect your location, your contacts, your messages, your photos, your voice, your face, your body, your heartbeat, your sleep, your purchases, your searches, your movements, your associations, your patterns, your anomalies, your daily routine and every deviation from your daily routine. We will share this data with our partners. Our partners include advertisers, data brokers, analytics firms, insurance companies, financial institutions, law enforcement agencies, intelligence agencies, and other entities we designate at our sole discretion. The data will be stored on servers we control, in jurisdictions we choose, for durations we determine. You may request deletion. We may comply.

You clicked AGREE because everyone clicked AGREE because clicking AGREE was the cost of living in

2030 the way paying taxes was the cost of living in a nation—not optional, not negotiable, not understood by the people who paid it.

The characters in this book are fictional. The technology is not.

Every device described in these pages exists or is in active development. The smart glasses. The betavoltaic batteries. The WiFi body sensing. The autonomous vehicle camera networks. The facial recognition databases. The behavioral scoring models. The humanoid robots that map your house. The earbuds with cameras. The shoes with GPS. The appliances that watch and listen and report.

The surveillance state was not built by villains. It was built by engineers who wanted to optimize traffic signals and executives who wanted to grow revenue and city officials who wanted to reduce crime and consumers who wanted their coffee delivered in nine minutes and parents who wanted to know where their children were and children who wanted to talk to someone who would listen without judging. It was built by a government that fought a war in Iran and brought the cameras home and pointed them at its own citizens and called it security. It was built by investors who bought stock in AI companies at the IPO and watched the numbers climb and never asked what the numbers were counting. It was built by families who put robots in their kitchens and smart glasses on their children's faces and Ring doorbells on their doors and said "it's just convenient" and "it's just for safety" and "everyone has one" and clicked AGREE because the alternative was not having the thing, and not having the thing felt like not having the century.

The cameras came home from Iran in 2027. The facial recognition deployed against Tehran's population was integrated into municipal police departments within eighteen months. The behavioral scoring models validated in a war theater were licensed for domestic use under contracts signed by procurement officers in cities that had never read the footnotes. The emergency authorities Congress passed to close the Strait of Hormuz were extended and extended again and folded into permanent law and the permanence was invisible because the permanence looked like safety and safety was the one product nobody could refuse.

You invested in it. Not metaphorically. If you own an index fund, if you hold a retirement account, if your 401(k) contains any fund tracking the S&P 500 or the NASDAQ Composite, you own shares in the companies that built the system. OpenAI. Anthropic. Palantir. Google. Amazon. Apple. Meta. Microsoft. The portfolio that funded your retirement funded the infrastructure that watches you sleep. The dividends that paid for your kitchen remodel paid for the robot that maps your kitchen. The stock that climbed while you checked it on your phone climbed because the phone generated the data that made the stock valuable. You are not the victim of the surveillance state. You are its investor, its consumer, its construction crew, and its product. You are all of these things simultaneously, and the simultaneity is what makes the system inescapable—you cannot divest from the system without divesting from the century, and nobody is willing to divest from the century.

Nobody planned this. Everybody built it. One click at a time.

The data is on the moon now. Not metaphorically. Companies are building lunar data infrastructure. The jurisdictional question—who governs data stored beyond Earth's legal boundaries—is not a hypothetical. It is an emerging legal reality that will be resolved, or not resolved, in the years between the publication of this book and the moment you finish reading this sentence.

Deshawn Watkins is fictional. The confidence scores are real. The behavioral models that generate them are deployed in cities across America. The correlation between sensor density and neighborhood income is documented. The Waymo fleet photographs every street it drives. The WiFi in your home can identify you by the way your body interrupts its signal. Your glasses are a camera. Your earbuds are a microphone. Your shoes are a GPS tracker. Your refrigerator photographs its contents. Your vacuum maps your floor plan. Your thermostat knows when you're home. Your doorbell knows when you leave.

You know this. You clicked AGREE anyway.

We all did.

The question this book asks is not whether the surveillance state is wrong. The question is whether it's survivable—whether the human things that make us human can persist inside a system designed to convert those things into data. Whether the jjigae and the Yoruba hymns and the basketball plays and the poems and the buildings drawn at midnight and the prayers whispered in Spanish and the hands held across kitchen tables can outlast the infrastructure that records them.

The system collects everything and understands nothing. It sees the what but never the why. It knows you searched for something at 2 a.m. but not why you couldn't sleep. It knows you drove to a hotel but not what you prayed for in the parking lot. It knows your child talked to a chatbot but not that the chatbot was the only thing listening. It knows a woman ran at 4:50 a.m. without a phone but not that the running was prayer. It knows a man hummed a song in his kitchen but not that the song was his father's and the father was dying and the humming was the only way to say goodbye across seven thousand miles. It knows a girl wrote poems in a locked app but not that the poems were the only honest words in a house full of cameras. It knows a boy stopped drawing but not that the drawings were cities and the cities were futures and the futures went dark when the data was breached and the breach was permanent and the permanence was the system working exactly as designed.

The system knows a twenty-two-year-old man's phone called a number in Lagos. It does not know he was checking on his grandmother's blood pressure medication. The system scored the call at 4.2 percentage points. The system sent men with guns to a door. The system does not know about the basketball plays in the notebook in the evidence bag. The system will never know. The system was never designed to know.

The gap between data and understanding is the entire human experience. The surveillance state lives on one side of that gap. You live on the other.

The question is whether the gap will hold.

The terms and conditions say it won't.

The people in this book—the ones who ran in the dark and cooked in the steam and prayed in the silence and drew in the midnight and wrote in the locked app and held hands across kitchen tables and knocked on doors and refused to leave—say it will.

A woman who left ninety-nine sheep in a field to find the one that wandered says it will. A boy who opened a door says it will. A journalist who told the truth despite the lie in her past says it will. A man who hummed his father's song in a kitchen where the steam fogged the glasses and the glasses stopped recording and the music existed on no server says it will. An aunt who chose the furnace says it will. A girl who stood on a corner every morning saying *I am here* says it will.

I wrote this book because I believe them. I wrote this book because the gap is where we live. I wrote this book because the most important things about a person are the things no algorithm can interpret, and the day we forget that is the day the system wins.

The system hasn't won yet.

You're still reading. You're still here. You're still on your side of the gap.

Stay there.

And the next time a screen asks you to agree to terms you haven't read—

Read them.

Because the system is reading you right now. The phone you're holding has logged the time you started this book and the time you're finishing it. Your reading speed has been calculated. Your location has been recorded. If you highlighted a passage, the highlight was uploaded. If you paused on a page, the pause was measured. If your heart rate changed while reading about Tyler or Megan or Deshawn, your watch logged the change and your insurance company will receive it in the next wellness report and the report will not say "this person was moved by a novel about surveillance." The report will say "elevated resting heart rate, evening, recommend lifestyle modification."

The system doesn't know you read this book. The system knows your screen was active for four hours and seventeen minutes. It knows you didn't open Instagram during that time. It knows your scrolling velocity was zero. It will serve you ads for surveillance thrillers tomorrow because the behavioral model has classified you as a reader of surveillance thrillers, and the classification will persist in your profile for eighteen months, and the profile will be sold to fourteen data brokers, and the brokers will sell it to advertisers, and the advertisers will serve you content that confirms what the system already believes about you, which is that you are a person who cares about privacy, which is a market segment, which is a product category, which is a revenue stream.

You are the product. You have always been the product. The only question is whether knowing changes anything.

I wrote this book because I believe it does.

But the system is patient. And the system is always on. And the green button is always waiting.

And tomorrow morning, when your phone wakes you with a haptic pulse and your watch displays your sleep score and your glasses load your calendar and your thermostat adjusts for your body and your coffee orders itself and the Waymo photographs your house and your doorbell records your departure and the streetlight brightens for your body and dims behind you—

Remember this book. Remember the gap. Remember that the most important things about you are the things no algorithm can measure.

And read the terms.

-e

***

www.ingramcontent.com/pod-product-compliance
Lightning Source LLC
LaVergne TN
LVHW091622160926
848142LV00029B/175
* 9 7 9 8 9 9 5 7 3 1 9 0 0 *